The Apple II Age

The Apple II Age

HOW THE COMPUTER BECAME PERSONAL

Laine Nooney

The University of Chicago Press
Chicago and London

The University of Chicago Press, Chicago 60637
The University of Chicago Press, Ltd., London

Published 2023
Printed in the United States of America

32 31 30 29 28 27 26 25 24 23 1 2 3 4 5

ISBN-13: 978-0-226-81652-4 (cloth)
ISBN-13: 978-0-226-81653-1 (e-book)
DOI: https://doi.org/10.7208/chicago/9780226816531.001.0001

Library of Congress Cataloging-in-Publication Data

Names: Nooney, Laine, author.
Title: The Apple II age : how the computer became personal / Laine Nooney.
Description: Chicago : The University of Chicago Press, 2023. | Includes bibliographical references and index.
Identifiers: LCCN 2022045947 | ISBN 9780226816524 (cloth) | ISBN 9780226816531 (ebook)
Subjects: LCSH: Apple II (Computer) | Microcomputers.
Classification: LCC QA76.8.A66 N66 2023 | DDC 005.265—dc23/eng/20221021
LC record available at https://lccn.loc.gov/2022045947

♾ This paper meets the requirements of ANSI/NISO Z39.48-1992 (Permanence of Paper).

In memory of Margot Comstock, the founder of *Softalk*.
1940–2022.

Her passion for the Apple II left behind the trace that made this book possible.

CONTENTS

Introduction

"Every once in a while," Steve Jobs announced, "a revolutionary product comes along that changes everything."

It was the morning of January 9, 2007, at the annual MacWorld convention in San Francisco, and Jobs, Apple Computer's CEO and visionary in chief, was angling to make history.[1] Attendees from across the creative sectors had camped outside the Moscone West convention center since nine o'clock the night before, had endured Coldplay and Gnarls Barkley as they stabled inside the auditorium awaiting Jobs's arrival, had dutifully, even passionately, cheered for Jobs's rundown of iPod Shuffle developments and iTunes sales, along with a pitch for Apple TV packed with clips from *Zoolander* and *Heroes*. And as everyone suspected, it all turned out to be one long tease for the only thing anyone remembers. Steve Jobs was unveiling the iPhone.

Pitched by Jobs as the convergence of three distinct technological traditions—the telephone, the portable music player, and internet communication devices—the iPhone needed a conceptual on-ramp to help people frame its "revolutionary" potential. So Jobs opened with a little Apple history.[2] Behind him on stage, a slide deck transitioned to a picture of the iconic 1984 Macintosh computer, followed by the 2001 iPod. The selections were deliberate: both had shifted Western norms about how computer technology fit into our lives, changed our gestures, altered our workflows. In Jobs's foreshortened timeline, the iPhone was the natural inheritor of this trajectory of innovation, a capstone technology in a chain of hardware evolution straddling the end of one century and the beginning of another. The iPhone promised to braid Apple's many trajectories into a single impulse: by drawing a direct path of influence from the Macintosh's graphical user interface to the dial of the iPod to the touchscreen of the iPhone, Jobs presented the development of all three as natural and inevitable, subtly suggesting a metaphysical inheritance between them that transcends their status as mere "products." This is company history told only through its virtues, a techno-secular *Creation of Adam* in which the first man reaches out with a mouse and God swipes back with a finger.

But here we should pause, and consider all the "history" that must be put aside for Jobs's version of events to hold together. Elided by necessity are all the Apple products that flopped or failed, the ideas that stayed on the drawing board, the incremental alterations intended to get Apple through its next quarter rather than catalyze a revolution. Think of the clamshell laptops, the Lisa workstations, the QuickTakes, Newtons, and Pippins. But there is one absence in particular that reveals more than the rest: the Apple II. (See pl. 1.) Released in 1977, seven years before the Macintosh, the Apple II would become one of the most iconic personal computing systems in the United States, defining the cutting edge of what one could *do* with a computer of one's own.

Like the founding of Apple Computer Company in 1976, the development of the Apple II is often presented as a collaboration between Steve Jobs and his lesser-known partner, Steve Wozniak. In truth, the product was almost entirely a feat of Wozniak's prodigious talent in electrical engineering—though it likely would never have come to market with the bang that it did had Jobs not been rapaciously dedi-

Scenes from the introduction of Steve Jobs's 2007 iPhone keynote, highlighting Apple's history of technological innovation with the Macintosh and the iPod. Screen captures by author. Video posted on YouTube by user Protectstar Inc., May 16, 2013: https://www.youtube.com/watch?v=VQKMoT-6XSg.

cated to commercializing Wozniak's talents. It was the Apple II, not the Macintosh, that made Apple Computer Company one of the most successful businesses of the early personal computing era, giving the corporation a foundation from which it would leverage its remarkable longevity. Yet the system cannot appear in Jobs's harried, abridged history. The Apple II renders too complicated the very mythology he was trying to enact through stagecraft.

To attend to such absences in Jobs's speech, to its false starts and

Steve Jobs stands dwarfed by his own history, speaking before a 1976 photo of himself and Apple cofounder, Steve Wozniak. In the historical image, Wozniak and Jobs pose with Apple Computer's first product, the Apple 1 microcomputer. Photograph of Jobs taken at the Apple iPad tablet launch, held at the Yerba Buena Center for the Arts Theater in San Francisco, California, January 27, 2010. Photograph by Tony Avelar, image courtesy Bloomberg via Getty Images.

the things left out, is to make the case for a history of computing that is more intricate than our everyday blend of hype and nostalgia. For many readers, the history of personal computing *is personal*. It is something witnessed but also felt, a nostalgic identification with a technological past that lives within. It comes alive at surprising moments: a familiar computer in the background of a television show, a YouTube video recording of the AOL dial-up sound, a 3.5-inch floppy disk found in a box of old files. These intimate memories of simpler technological times routinely form the ground from which personal computing circulates as a historical novelty. And beyond our individual reminiscences, personal computing's past is continually leveraged by those invested in using the past, like Jobs did, to barter for a particular kind of future.

To move beyond this frame is to suggest that perhaps computers themselves cannot tell their own history—and their most ardent prophets are not the sole keepers of their meaning. The emergence of what we would now call personal computing in the United States, during the 1970s and 1980s, is a wondrous mangle. Yet when we fixate on lauding heroes, heralding companies, or claiming "revolutions," we limit the breadth of our understanding of—and actual appreciation for—the impact of computer technology and the work of history itself. Naive fixations leave us open to charismatic manipulation (a particular talent of Jobs's) about the reason these technologies exist, the forces that brought them here, and the transformative authority they allegedly possess. If these histories have thus far gone unheard, we may do well to ask ourselves what has been gained (and who gained it) by their exclusion.

What we may discover is that the story of great men, with their great technologies, launching their great revolutions, is a story more for *them* (and their shareholders) than for *us*. The history of personal computing's emergence is as much social, geographic, cultural, and financial as it is technical or driven by human genius. It is a story of hardware and software; children and investment bankers; hard-core hobbyists, enthusiastic know-nothings, and resistant workers; the letter of the law and the spirit of the code; dramatic success and catalytic failure. It is a story of people, and of money: who had it, who wanted it, and how personal computing technology in the United States was positioned to make more of it.

This is the journey this book will take you on, using the Apple II, and the software designed for it, as a lens to magnify the conditions that first transformed the general-purpose computer into a consumer product in the United States. It is a story that begins in the mid-1970s and extends into the mid-1980s, a time during which there was a marked proliferation of microcomputer use among an uncertain American public. This was also a period of rapid industry formation, as overnight entrepreneurs hastily constructed a consumer computing supply chain where one had never previously existed. In other words, this moment in time marks the very beginning of individual computer *ownership*—technologically, economically, socially.

And we could have no better guide through this misunderstood and often misrepresented terrain than the Apple II's hardware and software. As the very thing Jobs chose not to talk about, the Apple II nonetheless documents, better than any other computer of its time, the transformation of computing from technical oddity to mass consumer good and thus cultural architecture.

* * *

Sixty years ago, no one had ever held a computer in their hands, because computers were the size of rooms, or refrigerators, or boxcars. Microprocessors, the technological basis for all consumer-grade computer technology—from personal computers to Game Boys to digital alarm clocks—didn't exist. Yet from the intersecting interests of dreamy counterculture futurists and rapacious venture capitalists, from American tech entrepreneurs who thought there just might be a small market for DIY computers among ham radio hobbyists to magazine editors who wanted to grow the subscriber base of their periodicals, entire consumer industries would rise over the 1970s and 1980s, organized around what would become known as the personal computer. Over the course of our own lives, these technologies have unfurled as icons of popular culture and mass consumption: slid into pockets, fitted to desks and cars, innocuous machines keeping track of all the money that makes the world go 'round while also serving up the casual distractions that have become a hallmark of leisure time in the Age of Information.

We couldn't overstate it if we wanted to: computing, and in particular the various forms of consumer computing, of personal computing, has reorganized everyday life. But in those small moments half a century ago, when someone first unboxed a computer on their kitchen table, many consumers struggled to imagine how a machine that looked like a cross between a television and a typewriter might expand their creative potential or their business efficiency. As for the industry, the companies serving up the dream of personal computing were unclear and unconsolidated, sometimes barely turning profits—even as they were being actively underwritten by the financial hopes of a flagging American industrial economy. And yet beyond the usual suspects of Steve Jobs or Bill Gates, we know almost nothing of how the personal computing industry grew to even half of its present scale, of how it industrialized, corporatized, capitalized.

As mentioned earlier, many of this book's readers will have been drawn to it by an internal sense of having watched the history of personal computing "happen" right before their eyes, and this is especially true for anyone who participated in the rise of these industries. But historical knowledge accumulates in other ways too: through YouTube videos or Netflix documentaries, films and television programs that use the 1970s and 1980s as a historical set piece, articles in *Wired* or *The Atlantic*, perhaps even a book or two, like Walter Isaacson's *The Innovators* or Steven Levy's *Hackers*. Histories of personal computing are also continually drawn upon by entrepreneurs, investors, thought leaders, and the like, often presented much the way Jobs did at the iPhone launch: discussions are highly selective and forever progressive. We've been told, over and over again, in countless forms and by myriad voices, that personal computing was, from the moment of its invention, instantly recognized as a revolutionary technology and eagerly taken up by the American public.

This is not true. The 1970s and 1980s were not a period of mass computer adoption in American households. Indeed, the bullish estimates of late 1970s and early 1980s investors, entrepreneurs, and futurists fell quite short of the heady predictions that the market would double annually year after year after year.[3] Personal ownership of computing systems barely touched double digits by the mid-1980s and, according to the US Department of Commerce, had reached only a third of US households by the late 1990s.[4] Computers pollinated more quickly in businesses and schools, American institutions more vulnerable to appeals about how a computer might improve workplace operations or global educational competitiveness. Yet on the domestic front, the truth is that personal computers were not ubiquitous American household technologies. And this was not simply an issue of price. The computer was not the television or the radio or the microwave. It was not obvious or easy to use, even in its more commercialized forms, nor did it solve a specific problem. Rather, the computer's greatest strength was also, at the moment of its commercial emergence, its greatest weakness: a computer was never anything more than what one made of it. Reluctance, ambivalence, confusion, and frustration were recurrent responses, even among those excited about computing's possibilities.

Yet how even a fraction of the American public came to be convinced to make something of the personal computer is not a simple tale of the ignorant masses recognizing the power and importance of the so-called computer revolution. Rather, it is a story of the tremendous *effort* undertaken to present computing as essential, helpful, safe—and personal. In order to do so, this emerging technology was quickly fitted to a variety of mainstream cultural and political norms. Personal computing was celebrated as a means to preserve American global economic leadership, a way to backtrack the rising deindustrialization of the 1970s and advance American entrepreneurialism. Industry stakeholders eagerly leveraged laws and lobbied the government to protect their financial interests, even at the cost of their consumers. In a surreal defiance of 1980s demographics that marked an unprecedented rise in divorce and single mothers, the personal computer was endlessly positioned as the ideal addition to the nuclear family. And the personal computer was heralded as a device that could save a cratering national educational system—all in the service of upholding the cause of American exceptionalism. These ideas were not manipulations or malignant growths that formed atop some purer essence of what personal computing once was, or could have been. They were motivations, there from the start, that shaped what people imagined personal computing should be, and would become.

The subtitle of this book, *How the Computer Became Personal*, begs a historical question: What *is* a personal computer? Today the term is pure generalism, usually referring to any type of desktop computer intended for individual use. Yet it is also quite a mutable concept. Someone might call their laptop a personal computer, and the label would not be incorrect. When pressed, my students might categorize all kinds of individual computational devices as "personal computers," including tablets and smart phones. What makes a computer "personal" today tends to be defined by a level of *user intimacy*: a personal computer has become a computer that stores, and allows you to manipulate, the content and metadata that is unique to you, whether notes, photos, numbers, or social connections. It is often assumed that such a device is your *personal possession*, or at least your personal responsibility in the case of desktops and laptops provided by your employer. Thus the "personal" computer is financially, legally, and culturally distinct from computing as large-scale infrastructure (embodied in mainframes and supercomputers) or network installation (whether at the scale of cloud computing servers or the kiosk of an ATM). Rarely stated but usually implied is also the *general-purpose* nature of a personal computer, meaning that it should be able to carry out a variety of different tasks and, ideally, be programmable by the user. In this sense, personal computers are distinct from dedicated machines like an Alexa, a Nintendo Switch, or an iPad, despite the ability of dedicated machines to run a wide variety of applications. "Personal computing" has become a vague container, a category of digital devices we own and use in everyday fashion, rather than a label that tells us much about those devices' histories.

This present-day conceptual ambiguity is an extension of the contested history of personal computers themselves. Academic accounts of personal computing, such as they exist, often take the invention of the graphical user interface (GUI), and the Macintosh especially, as a starting point for personal computing as a technological genre. Thus the historical enterprise focuses on the genealogy of the GUI beginning from computer science researchers like Doug Engelbart and Alan Kay to the early application of GUIs at XeroxPARC to the commercialization of the concept in the form of Apple Computer's Lisa and Macintosh.[5] Beyond this trajectory, numerous works tie the rise of personal computing to the complex social juxtaposition of the 1960s West Coast counterculture to an influential community of Silicon Valley computing technologists—a trajectory most prominently cited to Fred Turner's *From Counterculture to Cyberculture*, though more likely deriving from John Markoff's *What the Dormouse Said: How the Sixties Counterculture Shapes the Personal Computer Industry*.

These approaches, in particular, have ingrained an interpretation of early personal computing as a tool for liberation that was later co-opted and commercialized.

While these longer technical and cultural prehistories offer one thread we can follow in the history of personal computing, they are also selective, cutting out much engagement with personal computing's actual commercial origins. For those who would find the GUI a starting point, it bears reminding that the personal computing industry was nearly a decade old by 1984, when Apple Computer released the Macintosh. The platform benefited from a complete industry infrastructure already in place, from the standardization of developer-publisher relationships and royalty structures to established distribution and retailer networks to the formalization of software categories to a fully developed arm of journalism attending to all of it. Furthermore, neither the counterculture nor prior academic research trajectories like those of Engelbart and Kay were the dominant force in the *nationwide commercialization* of personal computing from the mid-1970s on. While some members of the techno-counterculture were key players within hobbyist communities, led computing education initiatives, developed their own products, or even became prominent speakers and futurists (figures such as Jim Warren, Bob Albrecht, Ted Nelson, and Lee Felsenstein come most readily to mind), they were a handful of the hundreds if not thousands of stakeholders shaping the development of personal computing at this time. Insofar as these stories have shrouded the origins of personal computing in a pre- or anticommercial mythos, we've paid less attention than we might to the flows of capital, there from the start, that allowed these industries to scale.

Given this book's interest in the personal computer as a commercial product, the history I trace here may initially seem similar to popular and journalistic histories some may be familiar with, which often tell this story as a chronology of specific technical objects and their inventors. These histories generally, as I do, situate the commercial origin of personal computing in 1975, with the invention and, more important, the advertisement of the Altair 8800, the first widely publicized general-purpose computer marketed to individuals. From there, the story typically moves to what is known as the 1977 Trinity, referring to the concurrent release of the first three mainstream consumer-grade desktop computers: RadioShack's TRS-80, the Commodore PET, and the Apple II. These histories then take off in many directions, according to the taste of the author, but typically maintain an emphasis on genius inventors, their landmark innovations, and the megabucks they made (or lost).

While this book's objects may map to many of these contours, its pathway through them diverges significantly from popular and journalistic history. For me, the Apple II is not the star of the show; it's the spotlight that illuminates the stage. As such, this is certainly *not* a story of how the Apple II created the conditions for the computer revolution. You'll find no delirious claims to firsthood here, no Hail Marys of historical relevance. Rather, this is a story about how the Apple II is an optimal historical object, a platform *through which* we can locate an account of the rise of personal computing in the United States that is both technical and cultural, economic and social, sufficiently broad and generalizable, yet nonetheless particular, special, specific. Interweaving the cultural and industrial perspectives, this book offers a deeper portrait than either approach could provide alone.

Rewiring our assumptions about personal computing, whether popular or academic, inevitably also requires that we address the language that we use to talk about it. So from here on, this book uses the term *microcomputer* rather than *personal computer* and talks of *microcomputing* instead of *personal computing*. From roughly the mid-1970s to the early 1980s, the term *personal computer* was just one of a host of phrases circulated by users, hobbyists, programmers, journalists, and marketers, including *appliance computer*, *individual computer*, *home computer*, *small systems*, or, as I prefer, *microcomputer*—the most generic term for any general-purpose computing system appropriate for individual use in terms of both size and cost.

Furthermore, the meaning of "personal computer" was not stable during this time. In 1981, IBM released its first microcomputer, the IBM PC, in an attempt to dominate the business market for such machines. At this moment, "personal computer" (abbreviated by IBM as PC) became associated with the market category of desktop computers intended for individual use in offices. Personal computers thus became opposed to *home computers*, the term for desktop computers intended for nonprofessional purposes, such as hobbyist programming, games, or basic household management applications.[6] These distinctions were not just semantic. Microcomputers designed for office use and microcomputers designed for home use were different types of microcomputers, separated by criteria like technical specifications, the use of a television as a monitor (more typical in home computers), the kind of software available, the level of product support supplied by the manufacturer, and price point. These divisions thus had implications for how, and for whom, these devices were designed, distributed, and sold, as well as for how industry analysts quantified the unit sales and revenue of microcomputing products and companies. It was only with the production of lower-cost PC-compatibles and clones during the mid-1980s that "personal computer" again began to encompass the entire category of desktop computers.

But I also embrace the term *microcomputer* to retain a sense of alien distance from these technologies. Rather than construct a tight continuity between the past and the present, I want to hold these technologies at bay, to insist that, as both concepts and products, they would not be wholly recognizable to us today and that we should be wary of seeing the present in every shadow of the past. In other words, it was not immediately self-evident what it might mean for a computer to *be* personal. Likewise, computers did not become personal on their own, through some sheer technological force of will. The idea of the personal was one part of a diffuse and centerless strategy pressed by those with social, ideological, and usually financial investments in the growth of the industry, a way of framing a deeply unfamiliar technology as something one should personally desire. In this sense, the oft-cited computer "revolution" was less of a revolution and more of an ongoing cycle of iterative justifications for why people needed computers at all.

* * *

But what is it about the Apple II that makes it an ideal object for traversing such complex history? While the platform was not singular or causal in spurring the rise of the microcomputing industry in the United States, the Apple II has a number of qualities that make it especially suitable to this book's purposes. Chief among these is the platform's versatility. In a moment when the consumer market for microcomputers was just beginning to turn from hard-core hobbyists to curious early adopters, the Apple II was engineered in such a way that it could be treated as an off-the-shelf microcomputer (no tools required) *and* as a sophisticated platform for hobbyists (you can hardware hack if you really want to). The Apple II's unique hi-res graphics mode, its eight expansion slots, and Apple's early-to-market floppy disk drive peripheral all contributed to producing a machine that was remarkably ambidextrous in a market encompassing the interests of both experienced hobbyists and technical novices. Consequently, the Apple II was widely considered one of the few microcomputers that straddled the home and workplace computing markets: robust enough for the office, exciting enough for games in the home, expandable enough to be a tinker toy in the garage.

Yet a history of the Apple II, as a piece of hardware, cannot actually tell us much about how microcomputing gained traction in homes, businesses, and schools. That requires understanding how *software* developed—and here again, Apple offers an exemplary case. Apple Computer supported a flourishing third-party software market during a time when software developers had to make critical decisions about which platforms to devote their energy and startup capital to. There was no "PC standard" or interoperability during this time. Apple II software wouldn't work on a Commodore, which wouldn't work on an Atari 800, a TRS-80, a TI-99, or anything else that wasn't the specific hardware it was designed for. These constraints meant early developers typically wrote their software for one or two specific platforms first, and sometimes exclusively. In turn, the volume and quality of software for a given platform was a significant consideration for early microcomputer consumers. By fostering a third-party software market, Apple set itself apart from its competitors as a development platform. For example, RadioShack barred third-party software for its TRS-80 from being sold in its storefronts, while Atari would not release the source code for the 8-bit series' graphics routines. Apple's robust system and hands-off approach to the creative impulses of Apple owners turned Apple programmers helped create products, and thus a market, for the very nontechnical customers Apple would need in order to dominate the industry.

By the end of 1983, the Apple II and IIe family had the largest library of programs of any microcomputer on the market—just over two thousand—meaning that its users could interact with the fullest range of possibilities in the microcomputing world. This gamut of software offers a glimpse of what users *did* with their personal computers, or perhaps more tellingly, what users *hoped* their computers might do. While not all products were successful, the period from the late 1970s to the mid-1980s was one of unusually industrious and experimental software production, as mom-and-pop development houses cast about trying to create software that could satisfy the question, "What is a computer even good for?" The fact that this brief era supplied such a remarkable range of answers—from presumably obvious contributions like spreadsheets, word processors, and games to remarkably niche artifacts such as recipe organizers, biorhythm charters, and sexual scenario generators—illustrates that, unlike popular accounts that would cast the inventors of these machines as prophets of the Information Age who simply externalized internal human desires, computing was an object of remarkable contestation, unclear utility, futurist fantasy, conservative imagination, and frequent aggravation for its users. Software is an essential part of this story because it was through software that the hypothetical use cases of the microcomputer materialized. Through software, consumers began to imagine themselves, and their lives, as available to enhancement through computing.

This mutually constitutive relationship between hardware and software guides the organization of this book. To lay essential groundwork, the book opens with a speedrun of US computing history from the 1950s to the 1970s, tailored to explain how something like a microcomputer even becomes possible. Then we slow down for a chapter, to focus in greater detail on the founding of Apple Computer, the invention of the Apple II, and the respective roles played by Steve Wozniak, Steve Jobs, and Apple's first major investor, Mike Markkula. Together, these chapters have a shared goal: to lay the necessary technical, social, and economic context for the rise of the American consumer software industry under the aegis of the Apple II.

Yet rather than treat the American consumer software industry as monolithic or centralized, the book splits off into five concurrent software histories, each tied to a specific kind of software categorization and told through the story of an individual software product:

- Business: *VisiCalc*
- Games: *Mystery House*
- Utilities: *Locksmith*
- Home: *The Print Shop*
- Education: *Snooper Troops*[7]

The gambit here is that the emergence of specific software categories is actually itself a history of what people imagined computers were for, how people used their computers, and how they imagined (or were asked to imagine) themselves as users.[8]

In each case, the purpose of various software categories started out fairly indeterminate in the mid- to late 1970s but quickly formalized into the mid-1980s through the tangled efforts of developers, publishers, journalists, investors, industry analysts, and users themselves. Tied up in these stories of economics and industry are people: the consumers whose preferences, desires, and needs emerge as aggregates tallied up in sales charts; the users who posed questions and expressed opinions in letters to the editor; the magazine owners, editors, and journalists who mediated consumer moods and ambitions; and, of course, the programmers, investors, marketers, and company founders who didn't just imagine a new world of software, but made it, financed it, and sold it.

By the mid-1980s, the dominant frameworks had taken shape and separate software user markets were firmly in place; they would not alter dramatically over the coming decades. This book finds its exit around the same time, largely due to these consolidating tendencies, though other considerations inform this closure. By the mid-1980s, the centrality of the Apple II as a development platform was waning, particularly as the technical specs of individual systems became more similar. In addition, new forms of hardware competition shortened the horizon of the Apple II as a leading machine. Apple's 1984 release of the Macintosh, driven by a graphical user interface and fully closed-off internal hardware, clearly flagged the very different future that Apple—and Steve Jobs specifically—saw for what might make computing "personal."[9] Moreover, the increased mainstreaming of the IBM PC and its MS-DOS operating system, along with the rise of cheap PC-compatible clones and workalikes from Compaq, Tandy, and Dell, created a new "standard" from which future computing infrastructures would be built.

* * *

In some ways this book may seem a mangle of contradictions. It claims to focus on a specific piece of hardware (the Apple II) but is more generally about software. It purports, in some fashion, to be a history of use but is perhaps more accurately a history of American industrial formation. Each chapter is unique to the category and case study it focuses on, yet there is a great deal of common terrain between them: corporate history, sales analysis, and attention to how various forms of privilege shape what kinds of opportunities are made available to what kinds of individuals. This book wants to focus on the Apple II industry, yet in many moments it exceeds that boundary when the specificity of platform is not helpful for understanding the larger machinations of the software market.

It is a story that must also move between inventors, consumers, and this emergent industry's role in a larger speculative economy. The history of technology is full of moments when the creative and technical problem-solving of an individual or group of individuals materially *matters*, altering the terrain of the technologically feasible. The early history of microcomputing is rife with wicked problems that some people were, for any number of reasons, better at solving than others, as well as moments when an individual's ability to read or anticipate a consumer desire transformed their range of opportunities. As for consumers themselves, I want to give sincere consideration to people's curiosity and willingness to purchase expensive software they had never seen, to read dense and incomprehensible manuals, to write letters to computer magazines begging for help, to bang their hands against keyboards, to break copyright law in order to make backup disks—all in the hope that microcomputing would prove itself worth the trouble.

Yet these moments never happen in a vacuum. It is no accident that so many of the individuals in this book are straight, middle- to upper-class, educated white men, operating in relatively supportive personal environments. This book takes extra care to consider the background, environment, and structural advantages of individual historical actors under examination, sometimes with surprising conclusions; for example, access to Harvard, rather than Silicon Valley, is a defining attribute for many of the entrepreneurs this book surveys. As such, this book does not capitulate in its certainty that most of what made personal computing "happen" was the financial interests of an elite investor class who were less interested in producing a social revolution than they were in securing their financial standing in the midst of the economic uncertainty of the 1970s and early 1980s.

These tensions, at times irresolvable in the project of historical storytelling, are what the Apple II Age *is*: not the story of a computer, but a roving tour of the American microcomputing milieu. I cannot imagine telling the story differently; refusing clean focus is central to the work. This is a harder historical task, less heroic, less obvious, but it also helps dismantle our sense that any technology, past or present, should be taken as inevitable, unchangeable, or apolitical. Understanding that computing has always been a story of contexts, rather than triumphs, broadens our horizons of interpretation in the present. So to understand all this, we need to stick with it. Embedded in the Apple II is a history at last ready to move beyond hagiographic accounts of the brilliant loner who engineered it (Steve Wozniak), the charismatic bully who designed it (Steve Jobs), and the savvy investor who financed it (Mike Markkula) and get its hands dirty with a greater historical challenge: understanding the microcomputer's impact on everything else.

1

Prehistories of the Personal

The Apple II came to market in 1977, a year when we can safely say the vast majority of Americans were *not* thinking about computing. Of course, this is not to suggest that computing had no impact on their lives. Mainframes and minicomputers had been automating everything from airline reservations to insurance claims for well over a decade.[1] The census was calculated by computer, buildings and automobiles were increasingly designed with the aid of computers, and universities and school districts were steadfastly acquiring their own computer networks. In some professions, the use of simplified, single-purpose computing terminals was common, especially in fields involving data entry or word processing. And at the consumer level, handheld calculators, retail UPC scanners, and arcade games were nearly ubiquitous. Computer technology was everywhere, even if it didn't yet come in the form of the "personal" computer.

Yet during the 1970s—a decade marked by the fluctuations of an international oil crisis and the unraveling of the Vietnam War, colored throughout by intense waves of recession, stagflation, and unemployment—everyday people were not enthusiastic about, or even terribly interested in, *owning* their own computer. A case in point: in 1976, David Ahl, editor of *Creative Computing*, arranged interviews with several hundred adults and young people to hear their thoughts on computing, surveying them on topics ranging from the impact of automation on American job prospects to whether computers would improve health care.[2] He found that at least a third of the interviewees lacked a basic understanding of how computers worked. Moreover, people were confused when asked what they might do with a computer in their home. Some people presumed the questions referred to handheld calculators or robots. As one interviewee responded, "Well, maybe I'll have [the computer] serve me martinis when I come home from work."[3]

The dawn of consumer microcomputing in the United States thus reflects a strange set of affairs: computing had come to the people, but most of the people didn't care. The broad adoption of microcomputing in homes, schools, and offices did not happen with the snap of a finger, and despite anecdotal claims, microcomputing wasn't widespread, at least in homes, for nearly two decades. While some historians have claimed that 1977 marked the "single moment when the personal computer arrived in the public consciousness," this framing elides much of what is most intricate about this technological era in US history: how much people struggled to imagine what they would do with a computer at all.[4]

To understand how the technological reality, economic form, and cultural category of microcomputing emerged in the 1970s, we must first familiarize ourselves with what computing was *prior* to this moment. Rather than imagine microcomputer development as a causal chronology spread out along a timeline, we must consider multiple historical trajectories within computing and electronics technology intertwining over time. Foremost among these are substantive transformations in *data processing and interaction* during the 1960s, encapsulated in the transition from batch-processing mainframes to time-shared, interactive minicomputing, which shifted the habits of computer use over the decade. Second is the phenomenon of *minia-*

turization, also accelerating across the 1960s and into the 1970s, in which tremendous financial incentives fueled research into the production of smaller, faster, and more stable computer componentry. I also trace the desire for *individual ownership* that suffused early homebrew microcomputing, as 1970s electronics hobbyists generated new possibilities for what they might do with computers they could own. Finally, I examine the imperatives of *profit-seeking*, in which both entrepreneurs and established electronics companies sought to meet perceived market demands for computer technology in the mid-1970s. By foregrounding these broader themes and phenomena, this account emphasizes habits, contexts, and practices rather than individual events or inventors. In doing so, this chapter parses what was actually important in the history of computing as it contributed to the development of the Apple II. As we will see, what was important wasn't always just technological concerns.

* * *

If people struggled to imagine what they might do with a computer, this was largely due to the inaccessibility of computing technology since its inception in the 1940s and its associated ties with the US government, military defense, and scientific research.[5] The computer's first mainstream televised appearance happened on November 4, 1952, when CBS brought a massive UNIVAC on set for election night coverage of the presidential race between Dwight D. Eisenhower and Adlai Stevenson. But the gimmick did little to normalize computing.[6] The UNIVAC towered over its commentators, blinked in incomprehensible patterns, refused to speak when spoken to. This alienating depiction of computing would dog fictional representations for years to come, from the romantic comedy *Desk Set* (1957) to the sci-fi epic *2001: A Space Odyssey* (1968) to the Cold War thriller *Colossus: The Forbin Project* (1970). (See pl. 2.) In each of these films, computers plot against their owners or threaten the autonomy of human intellect. Despite the prevalence of computing in nearly all major commercial industries and government institutions in the 1950s and 1960s, popular representations de-

The IBM 7094 Data Processing System, a mainframe computer released in the 1960s. The multiple control panels and more than a dozen tape reel storage units demonstrate the significant size and scale of mainframe installations. Image ca. 1965, courtesy the Computer History Museum.

picted it as ominous and threatening, and most people lacked the lived experience necessary to develop an alternative perspective.[7]

The computers depicted in cinema from this period are massive and foreboding, overtaking rooms, stretching the length of walls, bathing humans in the glow of their blinkenlights. These depictions are exaggerated but not wholly inaccurate, as CBS's stunt with the UNIVAC proved. Midcentury computing was defined by mainframes, which were massive (typically room-sized) centralized computing installations designed for large-scale institutional, governmental, or commercial data processing.[8] A single mainframe might serve the computational needs of hundreds of users dispersed across a company or even a geographic region.

Yet individual access was rare. Not only did few mainframes exist in the United States in the early 1960s (approximately six thousand across the entire country), but they were also exceedingly expensive.[9]

For example, the leasing costs for IBM's popular 360 mainframe series ranged from $3,000 to $138,000 *a month* in the late 1960s, depending on model and configuration.[10] This made computer time a commodity not to be wasted on less than expert users. Even people who wrote programs seldom had direct access to whatever mainframe computed their code. Instead, a program was inscribed onto punch cards (the dominant, paper-based data storage platform of the era) and turned over to an operator (often a woman), who served as intermediary between programmers and the sacrosanct machine.[11] The sheer expense of computing is the dominant reason that before the mid-1970s individual people did not own or even lease computers. Computers proliferated in businesses, universities, federal research labs, local governments, and public institutions, but they were not found in people's homes.

Starting in the late 1960s, however, the norms surrounding computer accessibility began shifting as a result of two technological developments: the rise of a new method of data processing, known as time-sharing, and the development of a new class of computing technology, the minicomputer. Time-sharing was a transformation in the way computers processed data, which in turn altered the material usability and interactive experience of computing power. Prior to time-sharing, mainframes processed data linearly: operators would take the punch card programs submitted by many different users, batch them, and process them as a stack. Known as batch processing, this method minimized the amount of time a computer sat idle but meant that programmers had to wait several hours for results. With time-sharing, however, the computer divided its processing capacities between multiple users across a network, switching its processing activities from one user to the next rapidly enough that human perception did not experience much lag.[12]

These technical transformations in data processing had sociocultural dimensions. In other words, the experience of interacting with the computer changed. As the media studies scholar Kevin Driscoll has described it, "In the ideal case, a time-sharing system gave multiple users the illusion of having exclusive access to a machine all their

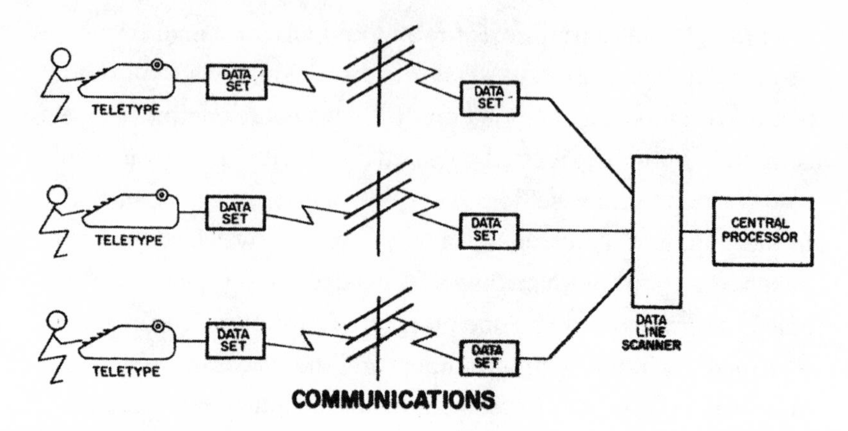

A diagram illustrating the distributed processing architecture of time-sharing. Multiple users connect to a central processor via teletype machines and a telecommunications network. Published in the *PDP10 Timesharing Handbook*, Digital Equipment Corporation, Maynard, MA (1970), 1–8. Image courtesy the Hewlett-Packard Company.

own."[13] With time-sharing, physical proximity to a computer was no longer a precondition for physical access. Distributed over a network, users did not have to be immediately local to the computer, or present with one another. Rather than have an operator as intermediary, users could access a computer's storage and processing capabilities directly using a terminal of some kind, typically a teletype machine, which was an integrated keyboard (for input) and printer (for output), though terminals with CRT monitors also existed.[14] Altogether, these intertwined changes in both the dynamics of data processing and the relations of distance between user and computer resulted in a sensation of immediate interactivity and individualized use—qualities that would later be understood as essential to the rise of microcomputing as "personal" computing.

Time-sharing first developed for mainframes in the early 1960s, but the practice truly took off with the emergence of a new class of computing systems called minicomputers. (See pl. 3.) Typically no bigger than a large refrigerator, minicomputers were smaller, cheaper, and better suited than mainframes to the data processing needs of universities, school districts, and midsized offices.[15] While every installation was unique, it was generally understood that a $20,000 minicomputer from 1970 had processing power equivalent to a $200,000 mainframe made just a few years earlier.[16] The economic gains were especially real: in the late 1960s, leasing a lower-end IBM mainframe cost $5,330 a month, yet the basic configuration of the Digital Equipment Corporation (DEC) PDP-8 could be purchased outright for only $18,000.[17]

Despite these comparative benefits, however, hardware manufacturers did not initially develop minicomputers to compete with mainframes; each class of computers was considered to have its own affordances and limitations. As Paul Ceruzzi notes, minicomputers were significant because they "opened up entirely new areas of application" and "introduced large groups of people—first engineers and scientists, later others—to direct interaction with computing machines."[18] The PDP-8, which would become one of the most popular minicomputers of the 1960s, eventually reached over fifty thousand installations. One of the machine's publicity photos even showcased it as theoretically portable, snugged in the backseat of a Volkswagen Bug.

Time-sharing and minicomputers profoundly transformed accessibility to computer processing power and in doing so, helped disseminate the general know-how required for computer programming. The computing historian Joy Lisi Rankin has argued that the prevalence and popularity of time-shared minicomputing, especially in educational settings, formed the basis for constructing "computing citizens," for whom accessible, user-oriented computing technology engendered "cooperation, inspiration, community, and communication."[19] This, in turn, laid a foundation for forms of personal and social computing that far precede our standard conception of these terms. In concert, time-sharing and minicomputing shifted the norms of interaction and access, introducing many of the intimate and individualized qualities we would later identify with the "personalization" of computing—qualities that would only amplify as computing devices grew smaller.

* * *

Achieving a mass-market, general-purpose computer small enough, inexpensive enough, and useful enough for individualized ownership hinged on an electronics innovation called the microprocessor (from which the "microcomputer" derives its name). For nontechnical users of the early microcomputer era, microprocessors were sometimes explained as the "heart" of the computer, the "brain," or just "where all the action is!"[20] While these terms are simplifications, they express something of the microprocessor's technical essence: the miniaturization of a computer's core functions, such that they could be centralized onto a discrete physical location often no larger than a postage stamp. (See pl. 4.)

The development of the microprocessor was part of a long trajectory of US government-backed research and investment in electronics miniaturization and semiconductor manufacturing dating to the 1950s, when the US military constituted 70 percent of the market for such goods.[21] When modern electronic computers were first developed in the 1940s, they were huge because they had to be. Early computers like the UNIVAC required many thousands of discrete electronics components—vacuum tubes in the UNIVAC's case, about five thousand of them, each roughly the size of a lightbulb.

Over the ensuing decades, research into miniaturization, materials science, and semiconductor manufacturing (much of it driven by the US government's demand for lighter, faster, and more reliable military and space technology) made possible dramatic reductions in the size of electronic circuitry. Bulky vacuum tubes in the UNIVAC (1951) gave way to boards of transistors and wire in the DEC PDP-1 (1959), which gave way to integrated circuits, also known as microchips, in machines like the IBM 360 (1964). The integrated circuit, first patented in 1959 at Texas Instruments, was a particularly critical development in this chronology, establishing a standard physical architecture for computing components that is used even today.[22]

By the 1960s, the integrated circuit had become the primary *technical* and *economic* unit of computer assembly. Miniaturization did not eradicate the market for mainframe computers—large-scale computing installations exist even today—but it did open up markets for smaller computing devices, such as minicomputers in the 1960s and consumer electronics such as portable calculators, digital watches, and video game consoles in the 1970s. The semiconductor manufacturers that made integrated circuits were driven by a hot and competitive market full of demand across the military and commercial sectors. The rivalry between these manufacturers further fueled research, as each sought the latest developments that might help them secure or maintain market share.[23] During this period of heady competition, the density of microchips doubled roughly every eighteen to twenty-four months as costs were driven exponentially downward (making good on projections made by Intel cofounder, Gordon Moore, and thus establishing the economic principles of "Moore's law").[24] By the late 1960s, microchips were dense enough that all of a computer's central processing functions could be contained on a single chip instead of having to be spread across multiple chips. In other words, the technological conditions of possibility for the microprocessor had been achieved.[25]

And the more chips manufacturers sold, the cheaper they were to make, driving prices down even as microchip complexity scaled upward. By 1975 it was possible to say that "the cost per component has in 10 years dropped by a factor of more than 100, from about $0.20 to a small fraction of a cent."[26] In consumer terms, a calculator that cost $100 in 1970 could be purchased for $10 by 1979.[27] This trajectory of innovation was not inevitable, but once started, the drive toward miniaturization was interpreted as both a cultural necessity and a technological inevitability.

The use value of such a technology, however, lay in its economies of scale. Before the microprocessor's development, microchips were typically custom designed for a specific device. For example, the first commercially released microprocessor, the Intel 4004, was originally designed by Intel as a cost-effective engineering solution for a Japanese electronic calculator company. The initial manufacturing schematics for the calculator had prescribed seven different integrated circuit chips, each one specialized, hardwired to a circuit board, and useful only for that device. By suggesting, instead, a single, general-purpose logic chip (in other words, a microprocessor) Intel could market that same chip for purposes beyond calculators. Thus, instead of producing tens of thousands of custom-designed chips for the manufacturing run of an individual calculator, Intel could mint hundreds of thousands of a single chip that could be used in any number of devices. Intel would grandly advertise the 4004 as "a microprogrammable computer on a chip!"—a clever bit of marketing lacquer, given that the 4004 needed three other integrated circuits to do anything useful.[28] But the sentiment was true in spirit if not in technical detail: what had once been spread throughout the metal chassis of a room-sized computer could now be held on the tip of a finger.

A progressivist history that imagines the rise of personal computing as obvious and inevitable might scold semiconductor and minicomputer companies like Intel, Xerox, and DEC for missing the opportunity to develop personal computers based on microprocessor technology. Ken Olsen, DEC's cofounder, never lived down his widely reported—and widely decontextualized—mid-1970s statement, "There is no reason for any individual to have a computer in his home."[29] Yet in the early 1970s these companies had good reason not to see opportunity in producing self-contained computing units for consumer purchase. Not only would such potential devices go against these companies' prevailing business models, service networks, and sales know-how, but the entire premise would have seemed technologically dead on arrival. The initial capabilities of a single "computer on a chip" were extremely limited, offering slower performance and a smaller memory address than the cheapest minicomputers.[30] DEC made a name for itself by designing robust computer architectures, customizing each chip for maximum efficiency; turning to a general-purpose microprocessor would have been seen as giving up what it was best at.[31] Even Intel's engineers presumed the primary market for microprocessors would be as control devices—the chips that regulate everything from microwaves to alarm clocks to automobiles.[32]

Thus Intel, like other microprocessor manufacturers, did not imagine consumers themselves would want access to such technology, given how difficult it was to program for. As a result of these presumptions, the earliest microprocessor distribution networks were not set up for single-unit or low-volume purchasing. As such, it was quite difficult for an individual to obtain a microprocessor in the early 1970s. Nonetheless, what manufacturing executives did not anticipate—and in some sense, had no reason to anticipate—was an entrepreneurial class of electronics hobbyists who got microprocessors circulating as products for individual use in computational experimentation.

* * *

Electronics hobbyism was an active leisure pursuit among a small but dedicated subculture of postwar American men, one that most prominently included ham and CB radio activities but also a broader range of electronics tinkering, repair, and kit building.[33] Such activities were part of a popular midcentury fascination with electronics, fed by industries with a concerted investment in consumer spending on novel goods and appliances. Millions of marketing dollars were burned popularizing a vision of the future in which electronics would "revolutionize everyday life."[34] Issues of the enthusiast magazine *Popular Electronics* from 1971 capture the vast range embodied by the hobbyist end of this cultural fixation: guides to constructing electronically activated locks, musical rhythm machines, solid-state lasers, electronic desktop calculators, and many more novelties of the so-called Electronic Age.[35]

While in the early 1970s it was still not economically practical to explore general-purpose computing at the hobbyist level, many of the devices showcased in *Popular Electronics* were computing adjacent insofar as they involved circuit board construction or even low-cost microchips.[36] For a community dedicated to rapaciously exploring personalized applications of electronics, there was tremendous enthusiasm surrounding the potential application of computing technology to hobbyist activities.

Electronics hobbyists established the conditions of possibility for the emergence of computer hobbyism. The demographics of this subculture therefore require consideration, insofar as they produced interlocking social constraints for who would and would not feel welcome within these communities. While midcentury electronics hobbyism in the United States has yet to be comprehensively analyzed, we can productively extrapolate from research focused on one of the most technically demanding electronics hobbies, ham radio. As the historian Kristen Haring has documented in her comprehensive exploration of ham radio enthusiasts, the ham radio community of the postwar period was "remarkably homogenous" in terms of race and gender, and its significant financial requirements further constrained it to the domain of educated white men.[37]

This social homogeneity was part of a longer trajectory in US history invested in the masculinization of electronics, particularly as a pathway to upward mobility for postwar middle-class white men. "High-tech employers," Haring writes, "endorsing the culture of ham radio, recruited hobbyists for the skills and traits developed through recreational tinkering. . . . There was a significant overlap between the groups that worked with electronics during the day for wages and in the evening for pleasure."[38] Ham radio hobbyists were also disproportionately represented in engineering and science occupations, meaning they typically had a practical exposure to computing that far outstripped that of an average American.[39] Ham radio enthusiasts would thus have been the most likely of the electronics enthusiasts to explore the possibilities of recreational microcomputing once it arrived to market. These enthusiasts also cultivated their own internal culture, organizing into local networks of clubs and user groups and sustaining communication and shared knowledge over larger geographies through newsletters and magazines. Through in-person activities and the circulation of hobbyist media, individuals united by technical interest cohered into subcultures of practice.

Thus we can recognize a set of mutually overlapping conditions surrounding work, leisure, gender, race, and class within the hobbyist communities that would first take an interest in computer ownership. These categories tended to reinforce each other, producing a closed subculture.[40] While the circulation of computer technology through hobbyist networks enabled new possibilities of access for those who were not a part of the military-industrial complex, cultural enclosure among hobbyists meant that the most immediate pathway of access to early recreational microcomputing continued to be closed to most women and people of color.[41] This is not to say that electronics hobbyists engaged in explicitly racist or sexist activity per se but to insist that their behavior was racialized and gendered insofar as participation in the hobbyist community hinged on certain socioeconomic advantages that intersected strongly with race and gender. Similarly, this is not to suggest that women or people of color had no participation in electronics or microcomputer hobbyism. Rather, it asks us to recognize the embedded historical challenges such individuals faced and the extent to which their relationships with these technologies were never going to be figured as "natural" in the way a white man's was.[42] In the long turn of history's screw, the historical demographics of electronics hobbyism set the stage for microcomputing's own emergent technical culture and independent hobbyist community—helping explain why early microcomputing was so readily identified as a white, male interest.

The overlap between general electronics hobbies and individual experimentation with computers increased in the mid-1970s. From 1974 on, enthusiast magazines like *Popular Electronics*, *QST*, and *Radio-Electronics* increasingly hosted scattered ads and articles for microprocessor-based computer kits and DIY projects, such as the Scelbi-8H and the Mark-8.[43] One of the earliest and most prominent computer hobbyist magazines, *BYTE*, was founded by a former ham radio magazine editor. In the introduction to the first issue, *BYTE*'s editor, Carl Helmers, explained that his earliest exposure to computers came by way of a ham radio friend.[44] Even the term *homebrew*, which would crop up among early microcomputing hobbyist communities, derives from amateur radio slang, describing home-built, noncommercial radio equipment.

The clearest evidence of the role electronics hobbyists played in the establishment of microcomputing as both a technological desire and an economic-industrial reality is found in the Altair 8800, which debuted on the cover of the January 1975 issue of *Popular Electronics*.[45] (See pl. 5.) Launching the issue with the blaring headline, "Project Breakthrough! World's First Minicomputer Kit to Rival Commercial

Models . . . ," *Popular Electronics*' breathless coverage of the Altair 8800 dominated the issue's editorial, promising that the "home computer age" was available for $397.[46] Manufactured by MITS, an electronic calculator company in Albuquerque, New Mexico, the machine looked nothing like what we would recognize as a computer today. It had no screen or keyboard, no mouse or graphical user interface, not even a command line. Under its industrial blue chassis, the Altair was full of boards and wires and chips. The computer's input and output were rudimentary in the extreme: programs were loaded into minimal memory using switches along the front panel, and output was registered in patterns of light that had to be translated out of machine code. The processing power was comparatively minuscule, and MITS wouldn't offer purchasable software until months later.

Yet the Altair 8800 struck a nerve in the body electric of a nationwide hobbyist community sutured together by clubs, user groups, and magazines like *Popular Electronics*. While the Altair was not the first microcomputer kit advertised in the hobbyist press, it *was* the first to sell more than a couple of hundred units, thus becoming a foundational object of hobbyist computing. The Altair fit into conceptual frameworks hobbyists were already familiar with: it was roughly no bigger than the radio equipment or small minicomputers such enthusiasts already knew; it was reasonably priced for its market, comparable to a ham radio setup; and it worked as a kit, something users could put together themselves. Though it couldn't do much, it could do more than the kit computers that came before it.[47]

In many ways, the Altair was what this community had been waiting for—with the emphasis on *waiting*. The Altair did not invent the idea of a computer one could personally own. Rather, it tapped into an ambient desire for ownership and individualized use that had been primed by other phenomena, including the hobbyist electronics industry and time-sharing. While electronics hobbyists nurtured active social networks, eagerly shared knowledge, and in some cases even believed in their moral right to free access across radio waves, telephone networks, or mainframe computing systems, at the foundation of most of these communitarian practices was an emphasis on the private ownership of technological appliances. A special freedom and self-determination was associated with having access to electronics that were typically sold as black-box goods or otherwise only available in regulated environments; many hobbyists felt their explorations liberated these technologies from bureaucratic control.[48] This privileged politics of possession was a centerpiece of electronics hobbyist vernacular: while the word *personal* never appears in the January 1975 issue of *Popular Electronics*, the phrase "your own" occurs more than twenty times. As much as DIY practices and homebrew electronics emphasized the noncommercial nature of hobbyist production, the purpose of the output was to create a technological world fashioned to one's own desires—a purpose the electronics hobbyist industry supported and benefited from by framing production in the context of consumption.[49]

While the Altair 8800 may not have impressed professional users of IBM mainframes or DEC minicomputers, a mid-1970s electronics hobbyist would have immediately grasped how dramatically different the Altair was from nearly every previous piece of consumer-grade or DIY electronics. Consumer electronics, whether store bought or hobbyist made, had always been single-purpose devices, meaning they could do nothing other than what they were built to do, whether to measure voltages, perform numerical calculations, or transmit radio signals. In computational terms, the hardware *was* the software. But what the mass-produced microprocessor made possible, and what the Altair 8800 delivered on, was an engineering distinction between hardware and software that was common in large systems, except now economically and technologically viable for small ones.

The gravity of this distinction is emphasized in *Popular Electronics*' January 1975 editorial, which focuses wholly on the Altair. As the editorial director Art Salsberg wrote, "Unlike a calculator[,] . . . [c]omputers can make logical *decisions* for an accounting system, navigation computer, timeshared computer, sophisticated intrusion system, and thousands of other applications. The 'power' of Altair 8800 is such that *it can handle many programs simultaneously*."[50] The word *simultaneously* has specific meaning in this context: not that the

Altair could run multiple programs at the same time (as we are accustomed to doing on computers today), but that it could simply *run different programs*. This distinction between hardware and software had important implications for the experience of computer ownership: a computer's applications were limited only by the material constraints of the hardware, meaning that as the size, shape, and data processing architecture of computers changed, so did imagined uses. What the small size and localized processing power of the Altair made possible was a redrawing of the boundaries of what a computer could do when applied to an individual's needs.

For some historians, the emergence of personal computers has been interpreted as a unilaterally consumptive practice marking a turn away from the shared, community-oriented practices of time-sharing—from personal computing to personal computers, as Joy Lisi Rankin frames it.[51] Yet not all computational activities made sense at a distance or on a collective network. The politics of use was not a binary between networked computing and individual computers. While time-sharing's emphasis on the network allowed for shared computing experiences, time-sharing's individualized, remote terminals also modeled a politics of personal access to computing power that hobbyist communities desired to see through to its furthest conclusion—just as they had with all preceding electronic technology, from radios to hi-fi sets to television sets. Hobbyists desired not just access to computing power in the general sense, but a computer appropriate to a context of use solely defined by its owner. This coalesced through emergent practices of computer use based on hobbyist philosophies of technology ownership more broadly: hobbyists had always owned their kits and radios and calculators, so why should a computer be any different? Furthermore, some part of that community had always turned a profit on these desires, and microcomputing would prove no different.

* * *

As the earliest of early adopters, US computer hobbyists had an important impact on the growth of domestic microcomputing. After all, it was their time, intellectual energy, and money that would assert at-home microcomputing as something worth doing. Hobbyists formed the first clubs and user groups essential to the production of shared knowledge about these extremely complex machines. Hobbyists founded the first newsletters, through which the earliest open-source programming languages circulated. Hobbyists would share software and hardware parts, from paper tape to microchips to wiring, in an effort to buoy and expedite the explorations of other hobbyists. And hobbyists would publish the first professional magazines, like *BYTE*, *Creative Computing*, and *Kilobaud*, which circulated know-how and news, in addition to providing a ready forum for advertisers.

Microcomputer hobbyist groups, especially those that formed amid the waning counterculture ethos of the 1970s San Francisco Bay Area, have been a subject of lively fascination for many popular books on computer history, from Steven Levy's *Hackers* to Walter Isaacson's *The Innovators*. It's easy to understand why. The Bay Area's dense crosshatching of technology industries and government research, incongruously operating in the same geographic and social milieu as an eclectic scene of techno-utopians and antiestablishment types, produced an idiosyncratic and highly localized technological culture that nurtured creative discovery in microcomputing.[52] Yet this scene was not monolithic in its political identity, nor was it representative of every computing hobbyist community across the country. As some hobbyists were tinkering, trading software, grokking machine code, and soldering circuits, others were taking out lines of credit, placing phone calls to semiconductor manufacturers, and poring over electronics surplus listings to see where they might get a good deal on wholesale aftermarket components. In some cases, these were the same hobbyists. Across the country, and in Silicon Valley too, some hobbyists organized their interests not just to make computers for themselves, but to sell them to others.[53]

MITS may have been strongest to market in this untested industry in 1975, but the microcomputer world would swell over the next three years, as electronics engineers and entrepreneurs experimented with the increasingly accessible microprocessors from Intel, National

Semiconductor, Texas Instruments, and MOS Technology. The accelerated production of numerous quasi-DIY microcomputers from 1975 to 1977 has been informally understood as the first wave, or "pioneering phase," of microcomputing.[54] This designation includes the Altair 8800 most prominently but also the IMSAI 8080, the Vector 1, the Heathkit H8, the North Star Horizon, the RGS 008A, and others. As a historical designation, these machines are generally united by their minimal, boxy appearance; their reliance on buttons, switches, or lights for input/output; and the high degree of technical knowledge necessary to build and operate them. While few of these first-wave microcomputers had the impact of the Altair, the point is less about the historical significance of any given machine and more about the fact that their collective proliferation constituted the recognizable beginnings of *an industry*.

The first-wave era of microcomputers points to what is always a volatile and fascinating moment in the intersecting histories of capitalism and technology: the leveraging of technological goods or services in the hope of generating an exponential accumulation of wealth. If the rise of these early microcomputers caught computing industry stalwarts like IBM, DEC, and Xerox unaware, it is also a testament to the remarkably abrupt bottom-up growth of microcomputing that companies like Heath and RCA, cornerstones of home electronics and amateur radio supplies, took several years to enter the market. The seemingly simple act of selling microcomputers through consumer-level supply chains required a matrix of skills few companies or individuals had in aggregate: not only the ability to engineer a working computer, but also sourcing parts, financing receivables, handling manufacturing, and managing shipping and returns, as well as marketing, advertising, and customer support.

Thus, while it may seem counterintuitive that the Altair was developed in Albuquerque rather than Silicon Valley, MITS's advantages are retrospectively clear, insofar as the company had the technical know-how, the consumer-level supply chain experience, and the internal responsiveness that came from being a barely midsize company. It's also worth noting that MITS was *desperate*: steep competition in the calculator market had left the company verging on bankruptcy by 1974. Faced with the risk of losing the business entirely, MITS cofounder, Ed Roberts, took a measured gamble that he could produce a fully functional general-purpose microcomputer at a cost low enough to turn a profit. It was fear, not fearlessness, that fueled the Altair's innovation. Yet Roberts's ability to pull off this scheme relied on his preexisting stature as a local business owner. MITS was $300,000 in debt in September 1974, when Roberts was in the thick of designing the Altair, and he had to take out a $65,000 line of credit to continue operating—an amount equal to about 21 percent of the company's outstanding debt.[55] The success of the Altair 8800 was not just the convergence of technical brilliance, supply chain experience, and institutional knowledge of the hobbyist market, but an intersectional array of financial privileges operating at the local level: the way men extend trust to other men over the certainty of a handshake.

Yet even with its optimal positioning, MITS was unprepared for the extraordinary demand the Altair received. By the end of May 1975, MITS was promoting the fact that it had shipped over 2,500 units to individual *and* industrial clients—far exceeding the 800 units total Roberts originally estimated.[56] The Altair's viability in industrial applications has been long overlooked by accounts that focus on homebrew hobbyists and the counterculture, ignoring the speed with which the Altair went from one man's techno-utopian dream toy to another man's embedded system in an office or manufacturing plant. Whereas the Altair and similar computers initially sold themselves as low-cost *minicomputers*, this language was largely abandoned by mid-1975 in both hobbyist and commercial-sector marketing.[57] *Microcomputer* became the terminology of choice, emphasizing these machines as a fundamentally different class of computer for a different class of user—hobbyists, sure, but also companies and organizations that didn't want the expense and hassle of a full-scale minicomputer setup.

This presentation of the microcomputer as a competitor to the minicomputer was explicit in MITS's marketing materials for the Altair 8800b, an expanded and improved model released in 1976 to shore up design flaws in the original Altair. "Imagine a microcomputer with all the design savvy, ruggedness, and sophistication of the best minicomputers," proclaims the Altair 8800b marketing brochure. "Imagine a microcomputer that will do everything a mini will do, only a fraction of the cost."[58] This revised marketing regime was complemented by an array of peripherals extending the Altair from a blue metal box to a customizable and intercompatible system, including 8-inch disk drives, a keyboard terminal, and a line printer. In little more than a year, the Altair had gone from *being* a minicomputer to functioning as a herald for a new class of computer companies angling to consume market share by *beating* the minicomputer.

The term *microcomputer* was essential to the formulation of an identity for these products. Systems like the Altair couldn't cannibalize the minicomputer market overnight, but these transformations of language cleared space for an emergent industry organized around the production of hardware, software, and peripherals for small systems computers using general-purpose microprocessors rather than customized chip architectures. As discussed previously in this chapter, the microcomputer was not without its conceptual kin. Timesharing had brought users to expect direct interaction with a computer. Personal calculators and other consumer electronics, whether homebrewed or store bought, helped normalize notions of direct ownership of electronics. "Office automation" systems from companies like Xerox and Wang typically modeled a one-terminal-per-worker logic, consistently depicted in marketing materials in countless images of women office workers dutifully engaged in data entry. In order to get a burgeoning microcomputer industry off the ground, entrepreneurs, marketers, and journalists borrowed from these preexisting conceptual frameworks while also asserting that the microcomputer was more than simply a diminution of the minicomputer or the transportation of work applications into the home. Rather, they insisted that these machines embodied a new terrain of computing distinct from what previous systems provided.

While the idea of selling or buying computer hardware made sense to even non-computationally inclined consumers, software had almost no precedent for marketing or pricing at a consumer level. Software had only existed as a discrete capital good since the mid- to late 1960s. Before then, computer users obtained software freely through user groups, directly through external software contractors for a flat fee, or bundled with the cost of a computer manufacturer's hardware as a "non-remunerative marketing service to users."[59] External software programming firms were the first to explore whether the software they created for one company might be directly marketable to another—a practice that only became cost-effective once computing installations became more standardized, more cross-compatible, and more widely circulated.[60] Price points for these earliest software products ranged anywhere from $2,400 to $30,000, depending on the scale of the software in question.[61] But in all instances, this was commercial software for commercial clients: software with long development cycles, software that required extensive documentation and product support, software integral to the day-to-day operations of the company.

Translating the corporate software product business model to an exploratory consumer market would prove challenging for early software entrepreneurs, who not only lacked precedents for appropriate pricing but also struggled to justify their financial self-interest to some segments of the computer hobbyist community. Even if computer hobbyists were aware that a commercial market existed for software sold to businesses, those with exposure to computing in institutional and educational contexts were accustomed to either developing code themselves or obtaining it freely through professional networks. Indeed, it was the sharing tendencies of these networks, and the people within them, that had spread and popularized programming languages like FORTRAN, COBOL, APL, and BASIC—languages historians consider core to the advancement of computing in the United States. As many computer hobbyists saw it, sharing these kinds of software utilities was an ethical obligation, part of both a generalized pursuit of knowledge and the dismantling of authoritarian control over information.

So imagine the computer hobbyists' curiosity when MITS's customer newsletter, *Computer Notes*, began listing prices for *Altair BASIC*, a BASIC programming language interpreter customized for the Altair, in spring 1975, just a few months after the computer's promotion on the cover of *Popular Electronics*.[62] Everyone knew this was precisely the kind of software that was necessary to turn MITS's blue box into a legitimate general-purpose computer, and BASIC was especially desirable given its relative ease of use and widespread popularity in educational settings. But at what cost? As a stand-alone software product delivered on either cassette or paper tape, *Altair BASIC* retailed from MITS for $500—more than the cost of the Altair itself. That cost dropped to just $75 when purchased with an Altair, an 8K memory board, and an interface board, but this bundling maneuver did little to dull the frustrations of computer hobbyists.[63] From a hobbyist's perspective, charging so much (or for some hobbyists, charging at all) for that programming language was an affront to the Altair's claims of affordability, to say nothing of the spirit of hobbyism more broadly.[64]

A variety of tactical responses to *Altair BASIC* unfurled over summer and fall 1975, highlighting the tenuous position consumer-grade software occupied in this burgeoning industry. In the homebrew spirit, some computer hobbyists elected to write their own BASIC and share it within their communities. The most notable example of this was TINY BASIC, a highly condensed and somewhat rudimentary form of BASIC published in the People's Computer Company's September 1975 newsletter.[65] Readers were encouraged to circulate, expand, and improve on the code base; TINY BASIC's authors described the software as a "participatory design project" shared among users. [66] Such activities illustrate the extent to which collaborative development and open access to information were the de facto ethos for many computer hobbyists. Yet this spirit of sharing did not limit itself to collaborative coding projects: *Altair BASIC* itself was also subject to extensive unlicensed reproduction—in other words, software piracy. Hobbyists, MITS discovered, were reproducing *Altair BASIC* within their own hobbyist user groups, duplicating it from one user to another as they would a piece of homebrewed software.[67] By the beginning of 1976, it was claimed only an estimated 10 percent of Altair owners had also purchased *Altair BASIC*.[68]

While these unlicensed duplications may have short-sheeted MITS's ability to leverage software as a supplemental revenue stream, the real loss of monetary potential was felt by *Altair BASIC*'s developers: a minuscule startup called Micro-Soft, helmed by the twenty-two-year-old Paul Allen and Bill Gates, who was all of nineteen at the time.[69] Gates and Allen had drummed up the idea of a BASIC interpreter for the Altair after seeing the January 1975 issue of *Popular Electronics*, then called up MITS and bluffed their way through a phone call with Ed Roberts.[70] They weren't the only computer hobbyists with this scheme: Roberts reported that he received many such calls and would give the contract to the first person who showed up in Albuquerque with a working program.[71] Developing nonstop for the next two months, Gates and Allen managed to be the first to get a tightly programmed and fairly sophisticated implementation of BASIC into Roberts's hands (in no small part thanks to the unsupervised access Gates had to a PDP-10 while enrolled as an undergraduate at Harvard).

The deal Gates and Allen eventually struck with MITS was multilayered. Foremost, MITS received exclusive worldwide rights and license to *Altair BASIC* for ten years, although Micro-Soft retained ownership of the software. In exchange, Micro-Soft received $3,000 up front and a per-copy royalty arrangement. Royalties on bundled sales ranged from $30 to $60, depending on whether the 4K, 8K, or extended version (bundled, those versions of BASIC cost the consumer $50, $75, or $150, respectively) was sold.[72] Royalties for the expensive stand-alone copies were split fifty-fifty between MITS and Micro-Soft, as were any sublicenses sold to other computer manufacturers. Since Gates and Allen retained ownership of the software, they could pursue sublicensing arrangements on MITS's behalf, taking a 50 percent cut of each flat-fee sublicense all while expanding their reputation as *the* company supplying microcomputer programming languages. Gates and Allen could have elected to sell *Altair BASIC* themselves, but they benefited by not having to engage with all the details MITS handled, including advertising and marketing, manufacturing, and sales and distribution. By keeping their mission focused on software and letting MITS handle the overhead, Micro-Soft avoided having to dramatically scale up its internal management or involve itself in extraneous parts of the supply chain.

In structuring compensation through a sliding scale of royalties, the arrangement between MITS and Micro-Soft established a new business model in the software industry. The benefit of a royalty arrangement was that it built on the *volume* of sales rather than a specific contract price.[73] In other words, economies of scale are significant to understanding why royalties would have been preferable. The decision hinged on the anticipation that the microcomputer itself was primed to sell en masse, catalyzing purchases of software through which Micro-Soft would recoup production costs, support further development, and produce profit. While there might have been tens of thousands of mainframes and minicomputers in the United States, the conceivable market for microcomputers was, hobbyists hoped, every household in America—*millions*. Royalties would allow Micro-Soft to grow alongside the market itself and tied the company's success to the proliferation of microcomputers generally rather than any specific machine. But in such a financial arrangement, every unlicensed copy of *Altair BASIC* registered as revenue lost for Micro-Soft. While tech startups rarely make for sympathetic actors, the consequences of piracy had clear economic outcomes: Micro-Soft's 1975 royalties' statement amounted to only $16,005, despite the fact that thousands of Altairs were sold.

MITS senior staff admonished users twice in the pages of *Computer Notes* over the course of 1975 but with little impact. Struggling to locate an effective strategy for addressing these duplication practices, Bill Gates made the hawkish decision to write directly to hobbyist communities themselves. Penning what he titled "An Open Letter to Hobbyists," Gates minced no words about what he thought was going on and the impact it would have on the microcomputing industry:

> The amount of royalties we have received from sales to hobbyists makes the time spent o[n] Altair BASIC worth less than $2 an hour.
>
> Why is this? As the majority of hobbyists must be aware, most of you steal your software. Hardware must be paid for, but software is something to share. Who cares if the people who worked on it get paid?
>
> Is this fair? One thing you don't do by stealing software is get back at MITS for some problem you may have had. . . . One thing you do do is prevent good software from being written. Who can afford to do professional work for nothing? What hobbyist can put 3-man years [*sic*] into programming, finding all bugs, documenting his product and distribute for free? . . . Most directly, the thing you do is theft.

Gates's letter circulated widely within hobbyist communities, being published and republished across dozens of user group newsletters and computer hobbyist magazines.[74]As the media studies scholar Kevin Driscoll notes, "Gates's letter unsettled [hobbyists'] idyll," asking them to consider how sustainable it was for software development to be driven by "the social and technical pleasures of solving a tricky problem, demonstrating a new bit of hardware, and competing with one another."[75] Most of the published responses to Gates's letter took offense at his accusations of thievery and typically tried to reframe the issue around the effectiveness of Micro-Soft's own economic arrangement with MITS or larger dynamics within the hobbyist software industry (including whether commercialization was viable at all) or else ignored the issue of monetary compensation altogether. "The archival record shows that some hobbyists maintained an ideological opposition to the commercialization of microcomputer software," writes Driscoll, "while others felt ambivalent. . . . For the ambivalent, Gates's letter assigned a new moral calculus to the exchange of software."[76] If many hard-core hobbyists remained at best unimpressed with and at worst ambivalent to Gates's argument, the imperatives of Micro-Soft's profit-seeking would find their outlet not in consumer-level unit sales of *Altair BASIC* but in the company's expansion into sublicenses, other programming languages, and other microcomputers. Machines that may have been competition to MITS's hardware were just clients-in-the-making to a software company like Micro-Soft.

What Micro-Soft had done with the Altair was just one model of an effort to extract profit from the labor of creating software. As much as some histories have emphasized the sharing ethos of hobbyists, microcomputing also clearly emerged from a crux of preexisting capitalist interests. These include the multimillion-dollar hobbyist electronics industry, a nearly three-decade history of contract programming in business sectors, and the preexisting service charges associated with much time-sharing access. Whether some hard-core hobbyists liked it or not, the capitalists weren't coming—they were already there.

2

Cultivating the Apple II

If his own accounts are to be believed, Steve Wozniak might have been the person *least* excited to be standing in a stranger's Menlo Park garage on the drizzly night of March 5, 1975. Wozniak had tagged along under the encouragement of his old friend and fellow Hewlett-Packard engineer Allen Baum. Both were under the impression that the thirty or so men who had gathered alongside them were there to discuss the finer points of making homebrew TV terminals for time-sharing and other applications. As it turned out, however, all people wanted to talk about was microprocessors.

Where Wozniak had landed that night turned out to be the first meeting of what became the Homebrew Computer Club, an event now considered momentous in the history of personal computing for kick-starting the West Coast's emerging microcomputing scene.[1] But that night, Wozniak was out of his element. "I felt so out of it—like, no, no, I am not of this world. . . . I don't belong here."[2] The meet-up was thrown together by two local computer hobbyists, who had pinned index cards to bulletin boards at a few key locations around the Valley and Berkeley, welcoming anyone eager to "exchange information, swap ideas, talk shop, help work on a project, whatever."[3] A star attraction of the night was the Altair 8800, which another local computer group had on loan from MITS. "They were throwing around words and terms I'd never heard before," Wozniak recalled, "talking about microprocessor chips like the Intel 8080, the Intel 8008, the 4004, I didn't even know what these things were."[4] It seems unusual, in retrospect: Wozniak was employed in the cutting edge of pocket calculator design, in the geographic heart of American electronics innovation, yet was unfamiliar with the new world of microprocessors—testament to both how good Wozniak was at burying his head in his work and how uncertain the microprocessor's initial diffusion into consumer electronics really was.

Yet while Wozniak, who reveled in being the most technically competent among his peers, may have felt momentarily alienated in the presence of so many people who knew more than he did, he was as familiar with the foundational principles scaffolding microprocessor technology as anyone in that garage if not more so.[5] His life, in that moment, is a case study on the privileges of place—of the economic, institutional, and intergenerational forces that allowed a region like Silicon Valley to flourish.

Born in 1950, Steve Wozniak spent his youngest years in Southern California, where his father, Jerry, worked a variety of engineering jobs. One of Steve's earliest memories was watching his father operate an oscilloscope, an electronics device used to display and analyze electric signals. In 1958, a job at Lockheed's new Missile and Space Division brought the Wozniak family north to Sunnyvale, in the crook of the San Francisco peninsula. Men relocated by the thousands to grasp a piece of the postwar suburban dream, and soon Sunnyvale, this "Boomtown on the Bay," teemed with engineers employed to keep the Space Race running.[6] Jerry Wozniak was just one of twenty thousand who would relocate to the area within a handful of years. They would find good company packed among the Santa Clara Valley's broader electronics industry, composed of growing companies like Varian, IBM's West Coast division, Fairchild Semiconductor,

Hewlett-Packard, and more—each one feasting on government contracts, diversifying through commercial markets, nurtured by research collaborations at Stanford. As the journalist Michael Moritz put it, "In the mid-sixties electronics was like hay fever. It was in the air, and the allergic caught it."[7]

In other words, there was nowhere in America better situated for a child to gain an unusually early, everyday exposure to electronics hobbyism than Sunnyvale in the 1960s. Wozniak learned first at the workbench of his father, who furnished his son with "classical electronics training" beginning as early as fourth grade.[8] Wozniak's memories of this time, documented in his autobiography, *iWoz*, establish the picture of a shy boy with an indefatigable thirst for understanding the mechanics of the world around him—a desire indulged by his father, who taught his son that "engineering was the highest level of importance you could reach in the world."[9]

Middle-class suburbs packed with engineers meant easy access to electronics parts, supportive educational environments, even a ham radio operator down the road who taught Wozniak the craft.[10] In 1964, Wozniak enrolled at Homestead High School in Cupertino, an institution described as "upper-middle-class, almost exclusively white and Asian[,] . . . a kind of sanctuary for the brilliantly dysfunctional" (we might pause here to wonder which modes of childhood "brilliance" are allowed to be dysfunctional).[11] There Wozniak enjoyed the guidance of John McCollum, a former Navy engineer and resident electronics teacher, who kept an expertly outfitted classroom cobbled together from the donations of the local electronics firms.[12] McCollum even granted Wozniak and Baum special dispensation to spend their Wednesday afternoons at the electronics firm GTE Sylvania, where they pestered the engineers with questions and learned how to program FORTRAN on the company's IBM 1130 mainframe.[13] On weekends, the Stanford Linear Accelerator Center (SLAC) Library was just a short drive away; no one ever questioned these quiet high schoolers, white and male, as they soaked up computer magazines and wandered the stacks (years later, a SLAC auditorium would become the permanent meeting place of the Homebrew Computer Club).[14]

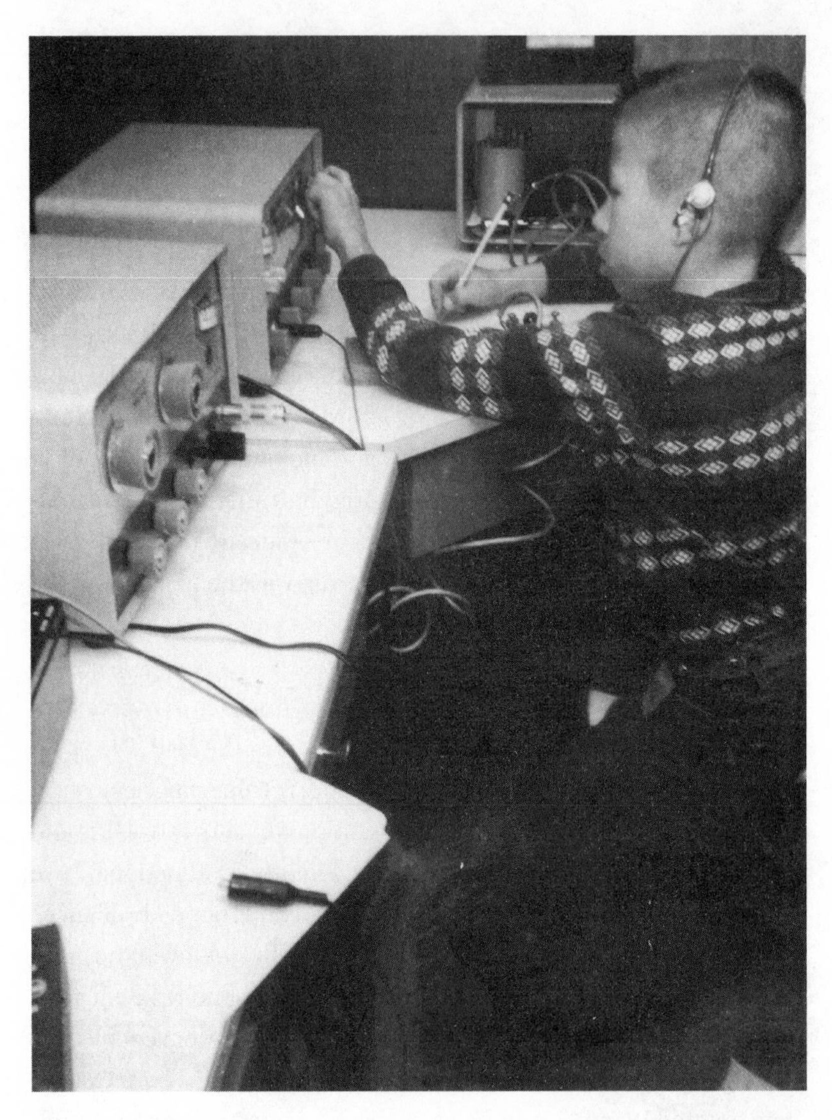

Steve Wozniak using a ham radio in his bedroom, at around eleven years old (ca. 1961). According to Wozniak, these were "the same desks [where] I taught myself to design computers." Image courtesy Steve Wozniak.

Thus the exposure that began with Wozniak's father was extended by the social bonds of his environment, as (typically male) teachers, neighbors, and family friends flexed their interpersonal networks and resources for his benefit.[15] This social ecosystem protected and amplified Wozniak's particular talents, offering him the space, time, and lack of accountability to other responsibilities necessary to exercise the diligent patience and intense focus essential for engineering. Wozniak's ability to move through this environment with support and safety dramatically affected his potential to follow his pleasure—a fact most readily exposed in the uninterrogated ease that flecks his biography, from "sneaking in" to Stanford's libraries to unobtrusively wandering the halls of a national electronics firm to leaving a fake bomb in his locker and getting away with nothing more than a night in juvenile detention. It was not by luck or chance that Wozniak thrived in this environment (as luck requires the even distribution of odds) but by the stacking of gendered, class, racial, and industrial advantages that eventually resulted in his exponential gains.

By late high school, then, Wozniak's accumulated technical know-how was tremendous. He was known to occupy himself by redesigning chip layouts for popular minicomputer systems, scoped in his head and drawn by hand on notepads.[16] Wozniak graduated from Homestead in 1968, with the counterculture blossoming all around, just up the peninsula in San Francisco and over the Bay at Berkeley. But that counterculture hardly earned a glance from the self-taught engineer. Wozniak scoped his dreams to the experiences with which he was most familiar; his greatest fantasy was to own enough chips to design his own minicomputer. Several years later, when Baum helped secure him employment at Hewlett-Packard, Wozniak believed he was landing his dream job: "For an engineer like me, there was no better place to work in the world."[17]

Wozniak's hobbyist energy was boundless. In the years between leaving for college in 1968 and his appearance at Homebrew in 1975, Wozniak compiled an accomplished resume of hobbyist creations, from a handheld device to interrupt television signals to a semifunctioning homebrew computer to a small box for hijacking phone lines (phone phreaking, as practitioners knew it) to a home *Pong* clone—among numerous other innovations.[18] Each of these inventions flexed technical competencies that Wozniak built on going forward and increased his familiarity with various electronics techniques, especially televisual output and economy of circuit board layout (the skill of using a minimum of integrated circuits to achieve a maximum of performance). These techniques, explored years before Homebrew, gave Wozniak the experiential knowledge to implement them in the Apple 1 and Apple II, thus lending those microcomputers their particular technical strengths.

Despite often being characterized as a loner, Wozniak routinely worked on these projects with friends who shared his interest in engineering. Wozniak built his first computer alongside neighborhood Homestead high school sophomore Bill Fernandez in 1971, while Wozniak was between years in college; they dubbed the twenty-chip machine the "Cream Soda Computer" in honor of the beverage that fueled its construction.[19] While designing the Cream Soda Computer, Fernandez introduced Wozniak to another Homestead high schooler, five years younger than Wozniak: Steve Jobs. Jobs, an intense and not particularly popular young man, found an easy friendship with the patient, forgiving Wozniak. The two remained close during the early 1970s despite their respective life transitions. Wozniak started at Berkeley that fall but dropped out after a year, did assembly-line work at a semiconductor components manufacturer, and eventually landed the Hewlett-Packard position. Jobs flitted about Reed College in Portland, occupied himself at a nearby commune, had a stint working at the arcade game manufacturer Atari, pursued Zen meditation and purification diets, and eventually traveled India, seeking spiritual enlightenment, for much of 1974.

But in their moments of intersection, Wozniak and Jobs pursued projects together—Wozniak as chief engineer, Jobs typically hustling for parts and sales. The first collaboration began shortly after they met in 1971, on the infamous "blue box." Blue boxes were devices for phone phreaking—the practice of using high-pitched tones to manipulate transmissions within a phone network, typically for the purpose of exploring the system or not paying for long-distance calls. After several attempts, the pair eventually fashioned a digital blue box running off a 9-volt battery, all of Wozniak's design.[20] The boxes made for excellent demonstrations with friends and provided a platform for countless pranks; Wozniak managed to call the Vatican, impersonating US Secretary of State Henry Kissinger (though he couldn't get the pontiff on the phone).[21] But where Wozniak saw fun and games, Jobs saw profit potential. Convincing Wozniak they could turn their blue box into a small business, Jobs sourced parts for casings and keypads, figured out the pricing strategy, then went dorm-room-to-dorm-room with Wozniak, selling the device in Berkeley's male residence halls (priced at $150, or roughly $950 in 2020 money, this blue box was a toy for Berkeley's richest boys).[22]

Jobs and Wozniak would again enter into casual business a few years later, in the early months of 1975, just before the first Homebrew meeting. Jobs was back at the Atari headquarters in Sunnyvale, having concluded his seven-month tour of India and a stint in primal scream therapy back near his Oregon commune.[23] Jobs was notably incapable of getting along with other engineers, having a habit of "regularly [telling] a lot of the other guys they were dumb shits," as the Atari founder Nolan Bushnell recalls.[24] One day, Jobs took a pitch: to design a single-player ball-bouncing game, a sort of riff on the original two-player *Pong*, with a bonus for using as few chips as possible.[25] No one thought Jobs had the technical chops for the gig, but as the mythology goes, Atari's engineering leadership knew that Wozniak was doing the work (Wozniak was a frequent visitor at Atari while Jobs worked there, known for spending lots of time hanging out on the *Gran Trak* machine).[26] Jobs promised the accommodating Wozniak a $700 revenue split if his friend could design the entire circuit layout in just four days—at a time when most new designs took three to five months to develop.[27] Wozniak's design was classic Wozniak: tightly scoped, elegantly engineered, utilizing a mere forty-some integrated circuits when games typically used over a hundred.[28] It was also so complex that there was no way to cheaply manufacture it. Eventually the whole project had to be redesigned, and it was released over a year later as *Breakout*.[29]

Both of these pre-Apple collaborations evidence the emerging shape of Jobs's and Wozniak's future entrepreneurial dynamic. The blue box episode established a pattern of Jobs leveraging Wozniak's technical skills toward shared moneymaking and Wozniak going along with it, mostly for the lark of it all. In financial matters, Wozniak was an even dealer: though Jobs stepped away from the blue box partnership before Wozniak did, Wozniak continued to split the earnings fifty-fifty.[30] But small details in the Atari episode reveal Jobs's fundamental orientation to self-dealing. The crushing four-day deadline was Jobs's invention, not Atari's (so crushing, by one report, it resulted in both men contracting mononucleosis); Jobs wanted the money sooner rather than later so he could return to his Oregon commune.[31]

But the promised revenue split has also become a key part of Apple apocrypha. The base fee for the project was $700, but by most accounts, Jobs never divulged to Wozniak that Nolan Bushnell also offered a bonus for every chip saved, which is alleged to have totaled $5,000 (roughly $24,000 today).[32] When asked about the matter directly by his biographer, Walter Isaacson, Jobs did what he had done since the story had surfaced thirty-five years earlier: he denied it. "I don't know where that allegation comes from. I gave him half the money I ever got. That's how I've always been with Woz. I mean, Woz stopped working in 1978. He never did one ounce of work after 1978. And yet he got exactly the same shares of Apple stock that I did."[33]

Such vainglorious accounting was a defining trait of Jobs, at any age. Born in 1955, Steven Jobs was the adopted son of Paul and Clara Jobs, a pair of working-class baby boomers who had settled in the San Francisco peninsula. Paul had a head for mechanics, fixing up old automobiles in his free time while bouncing between jobs as a machinist or doing the shadow labor of the banking industries—collecting bad debts, picking locks to repossess cars.[34] Paul often took his son along with him when he went haggling for used auto parts; numerous chroniclers have pointed to this as an early influence on Steve's disposition toward aggressive deal-making. But both parents believed in the value of sacrificing for their children. Steve's mother took babysitting shifts so they could afford to enroll their son in swimming lessons; they moved from Mountain View to Los Altos to put him in a better school district; and they supported their son's insistence on attending Reed College, a liberal arts institution that was far beyond their financial means.

If Wozniak was different from his peers in a quiet, unbothered way, the young Steve Jobs was known to make a showy insistence of himself as a free spirit and a freethinker.[35] While Jobs, too, was typically cast as a loner, many of his iconoclastic behaviors during the 1970s involved taking up space in ways that affirmed a sense of individualism: refusing to bathe, exploring a series of increasingly austere diets, upsetting others with purposefully confrontational conversations. He may have adopted the techniques of personal enlightenment—chanting with Hare Krishnas, dropping acid, practicing Zen before heading to work—but his particular embodiment of the counterculture ethos wasn't centered on a resistance to authority in a political sense. Rather, Jobs practiced a highly individualized mode of cultural revolt: a refusal to be defined in the terms of others.

What saved Jobs was a kind of fixed curiosity and blinkered faith he could extract from people, both those who worked for him and those in the position to do him favors. He was, by all accounts, a man of profound energy and intensity, his difficult demeanor punctuated just enough by moments of generosity and care. In direct conversation, he honed a charismatic capacity to direct his fullest attention to someone, exercising what one Apple employee, Andy Hertzfeld, called a "reality distortion field": "In his presence, reality is malleable. He can convince anyone of practically anything."[36] Like Wozniak and his social surround of engineers who wanted to see their craft thrive in the next generation, Jobs had no lack of people who not just tolerated him, but enabled him. The more abrasive elements of his personality were given a pass, especially by other entrepreneurial men, who saw in Jobs someone who might help make their own dreams of a computer revolution come true—even as he broke deals and willed other people into picking up the tab.[37] None of it would have gotten him far, however, without the backing of Wozniak's crafty machines.

* * *

Despite feeling alienated by all the microprocessor talk, Wozniak waited out the meeting. Sometime before it concluded, a fellow attendee passed out the schematics for an Intel 8008 microprocessor clone. Always curious about electrical engineering, Wozniak took the documentation home to examine later that evening. It was then, Wozniak claims, he began to grasp what microprocessors were:

> I knew exactly what these instructions meant. . . . I realized that all those minicomputers I designed on paper were pretty much just like this one. Only now all the CPU parts were on one chip, instead of a bunch of chips, and it was a microprocessor. . . . Then I realized that the Altair was . . . exactly like the Cream Soda Computer I designed five years before.[38]

In that moment, multiple strands of Wozniak's eclectic and wide-ranging electronics education twined together. His years spent studying and redesigning minicomputers on paper meant he had an extremely acute understanding of what microprocessor technology did under the hood (or inside the silicon, as it were). Like others of his cultural moment, he readily grasped the microprocessor's technological and economic affordances and recognized that his teenage desire—a computer that belonged *to him*—was possible. According to Wozniak, "It was that very night that I started to sketch out on paper what would later come to be known as the Apple I."[39]

Yet rather than start with a microprocessor and design outward, Wozniak's inclination was to implement the microprocessor inside a device he had already built, a TV terminal. TV terminals were, after all, the reason he had shown up that night.[40] In the engineering-centric social surround that constituted mid-1970s Silicon Valley, TV terminals were a popular part of the hobbyist repertoire, technological siblings to the commercial teletype machine discussed in the previous chapter. (See pl. 6.) Converging a keyboard and a CRT monitor, TV terminals contained just enough digital circuitry to connect a user, via phone lines, to a time-sharing service or other network. In other words, they were input/output peripherals, providing a centralized station for sending and receiving data but not carrying much computational power of their own.[41] Unlike expensive teletypes, however, a TV terminal could be fashioned at a hobbyist's workbench rather cheaply: Wozniak built his with a Sears black-and-white television and a $60 typewriter keyboard.[42]

But whereas time-sharing was about accessing computer power *through* peripherals and a network, a microprocessor allowed the computer *to come to* the TV terminal. Wozniak's design prototype thus treated keyboard input and screen output as essential components of computing rather than peripherals. In other words, Wozniak did not design the Apple 1 to use a keyboard and monitor as a way of innovating on the Altair. Rather, his thought was to use the microprocessor to draw the computer closer to the TV terminal.[43] While the product Wozniak eventually produced, the Apple 1, looks even shabbier than the Altair—given that it was just a free-floating circuit board without chassis, switches, or lights—its built-in connectors for television and keyboard afforded a kind of usability not previously possible for hobbyists. Wozniak's technical additions expressed a shift in how computers might be used and, inevitably, who they might even be for.

Wozniak wasn't alone among his peers in marrying the older technology of a video terminal with a microprocessor-based circuit board. The Homebrew stalwart Lee Felsenstein was developing a TV terminal–based microcomputer, the SOL-20, roughly around the same time as Wozniak, and video output was a frequent topic in the Homebrew scene's lectures and newsletters. What is significant here is not that the Apple 1 was first to market as a computer toting onboard video output (though it certainly was) but that the centrality of video output at this moment in time productively complicates any simplistic, linear account of computer history. The Altair did not "inspire" the Apple 1 through genealogical cause and effect. Rather, Wozniak's creation doubles back on computer history, goes around and behind it, suggesting that personal computing was not a forward march of "progress" but instead a mangle of media histories cross-wired between computing, networking, and television.

Wozniak began designing his homebrew microcomputer in March 1975, but he wouldn't finalize the design until September of that year, when he was finally able to source a new low-cost microprocessor, the MOS 6502.[44] The Homebrew Computer Club, which had grown from several dozen men gathered in a garage to a meeting of over a hundred held for three hours every other week in a Stanford auditorium, was an important context for Wozniak's ongoing production. By attending Homebrew meetings, Wozniak was no longer in the dark on matters like microprocessors. The vibrant collective energy these meetings attracted made them a first stop for anyone presenting microcomputing innovations or cranking on tough problems. It also became common practice for hobbyists to set up card tables in the hallway outside the auditorium to showcase their work—meaning the Homebrew Computer Club gave Wozniak a venue for gathering feedback and positive reinforcement about his incremental development. In taking his microcomputer to Homebrew, Wozniak had no commercial agenda. In fact, once he had a working prototype, he freely distributed his schematics in the altruistic hope that others would be inspired to build their own machines.[45]

What he began demonstrating at the Homebrew Computer Club was a smartly designed piece of hardware. In addition to the advantages of televisual display and keyboard input, his homemade microcomputer had the capacity to hold up to 8K of random-access memory, or RAM (making it equal to an Altair with moderate memory add-ons), and a software program permanently stored on a read-only memory (ROM) microchip, allowing the computer to immediately accept keyboard input when started.[46] And all of this was contained on a single circuit board.[47] Yet despite the cleverness of scale and its built-in convergence of monitor and keyboard, reception at Homebrew was tepid.[48] Because there was no interoperability between microprocessors, most of the Homebrew hobbyists were interested in computers based around microprocessors they were already familiar with, like the Intel 8080 driving the Altair. In other words, 6502-based machines were in the extreme minority of Homebrew projects.[49]

The impetus for commercializing Wozniak's computer came, not unexpectedly, from his friend Steve Jobs. Jobs had little interest in the technical minutiae that dominated Homebrew, but he attended a handful of times to help Wozniak show off his microcomputer. In doing so, he kept a careful eye on Wozniak's development process. By early 1976, Jobs began leaning on Wozniak to start another informal business venture—this time to manufacture a small run of Wozniak's circuit board design for sale to local hobbyists and retailers.[50] The pair's intent was not to produce a finished product but to sell just the circuit board for $50, leaving hobbyists to complete the computer themselves by buying the necessary chips and self-assembling. The plan was modest, but it had to be—neither had the means for anything more extravagant.

The pair moved quickly, pulling together the skeleton of a business over the early months of 1976. For the $1,300 startup capital necessary to lay out a producible circuit board design and get it manufactured, the two hawked their most valuable possessions: Jobs sold his Volkswagen Bus and Wozniak, his HP-65 pocket calculator.[51] Shortly thereafter, they settled on the name Apple Computer Company, an homage to the fruit Jobs had harvested during all the months he had spent working on the Oregon commune. Neither thought it was brilliant, but "Apple" stuck for lack of a better option.[52] And at some point, Wozniak took his microcomputer to Hewlett-Packard and offered them rights of first refusal for a design he had frequently fiddled with on company time, using company resources. Seeing no immediate appeal in the consumer-facing microcomputer market, Hewlett-Packard refused.[53] Jobs threw a mandatory business announcement in the *San Jose Mercury*, and on April 1, 1976, they signed the partnership agreement that brought Apple Computer Company into existence.[54] Their first circuit boards were manufactured with the designation "Apple Computer 1" at the left center edge.

* * *

Apple Computer Company might very likely have stayed a small DIY microcomputing operation if not for the financial intervention of several local entrepreneurs and investors, who saw good work in Wozniak's computer and a relentlessness in Jobs that spoke of the potential for bigger things to come. But beyond the appeal of Wozniak's tech or Jobs's personality, 1976 was also a moment when those close to the technology industries were looking to speculate on where the future was going. Nolan Bushnell at Atari had already proven that new consumer electronics industries could blossom nearly overnight. Simultaneously, government spending in the semiconductor industry was backsliding from its postwar highs. Capital needed other outlets for investment, and the slumped economy of the 1970s meant marketers were eager to find a different class of products that could appeal to a more precarious generation of consumers. As the so-called knowledge economy marched hand in hand with new tides of deindustrialization, some investors sought to redirect their money to hawk a future in which computing would be the new consumptive practice of the Information Age.

Apple's first financial opportunity came through a Homebrew Computer Club connection: a local entrepreneur, Paul Terrell, who owned Byte Shop, a newly established hobbyist computing retailer on El Camino Real. Terrell had seen the Apple 1 at a Homebrew meeting and was looking to expand his store's offerings.[55] But Terrell wasn't interested in retailing a raw circuit board that hobbyists had to finish themselves. As one of the only microcomputing retailers in the area, Terrell had a sense that the commercial market for microcomputers outstripped the narrow confines of Homebrew. What the customers coming into Byte Shop wanted was computing power without the labor of assembly—part of the shift in who computers were *for*, away from more technically minded hobbyists and toward "users." So Terrell made Jobs an extravagant offer: cash on delivery for fifty finished Apple 1 units at $500 each. In the estimation of Michael Moritz, "Terrell's order entirely changed the scale and scope of the enterprise. The size of the business had expanded tenfold and instead of contemplating costs around $2,500 for 100 printed circuit boards, Jobs and Wozniak were looking at a bill of around $25,000 to cover the costs of 100 fully assembled machines."[56]

What Terrell got, however, were not "fully assembled machines." Jobs was pulling the entire operation together on a shoestring. Unable to get a loan from a local bank, Jobs scraped by with one-month credit from an electronics supplier in Palo Alto, making thirty days the slim window he and Wozniak had to fulfill Terrell's order before interest would accrue with their supplier. They set up their assembly line in a spare bedroom of Jobs's parents' home, where Jobs paid his pregnant sister Patty $1 a unit to stuff chips into the empty circuit boards while watching soap operas. (See pl. 7.) What Jobs dumped off with Terrell was just the assembled printed circuit boards, lacking televisions, keyboards, casing, and any kind of onboard programming language. Yet despite not fulfilling his end of the bargain, Jobs walked away with the money Terrell had promised.

Terrell's purchase paid off the debt to the electronics supplier, with enough components left over to manufacture another fifty boards. Wozniak and Jobs produced those Apple 1s that summer—though Jobs's parents insisted Apple's operations relocate to the garage. They retailed the machine at the price point of $666.66 because they found it amusing (neither realized 666 had satanic connotations, amplified by the June 1976 release of *The Omen*). Jobs got a PO box and answering service for Apple to make the business appear more legitimate, and then he set about marketing his company's product.[57] Wozniak, for his part, was putting some incremental improvements on the Apple 1, including a cassette interface adapter, which would allow for cassette-based data storage, and a version of BASIC customized to the 6502 microprocessor.[58]

But despite these efforts, the Apple 1 didn't sell much during the remainder of 1976. With the rising popularity of the IMSAI 8080, an Altair clone released late in 1975, the hobbyist microcomputing world was increasingly consolidating around the Intel 8080 microprocessor, as well as the S-100 serial bus that had quickly become the data transfer standard for 8080-based machines. Hobbyists who purchased an Apple 1 would find themselves with an efficient and usable microcomputer but one unfortunately lacking available third-party software or peripherals. By most estimations, Apple was "too small, too fragile, and too eccentric to be taken seriously[;] . . . [it was] an unconventional local curiosity."[59]

The economic viability of the operation mattered little to Wozniak,

Apple Introduces the First Low Cost Microcomputer System with a Video Terminal and 8K Bytes of RAM on a Single PC Card.

The Apple Computer. A truly complete microcomputer system on a single PC board. Based on the MOS Technology 6502 microprocessor, the Apple also has a built-in video terminal and sockets for 8K bytes of on-board RAM memory. With the addition of a keyboard and video monitor, you'll have an extremely powerful computer system that can be used for anything from developing programs to playing games or running BASIC.

Combining the computer, video terminal and dynamic memory on a single board has resulted in a large reduction in chip count, which means more reliability and lowered cost. Since the Apple comes fully assembled, tested & burned-in and has a complete power supply on-board, initial set-up is essentially "hassle free" and you can be running within minutes. At $666.66 (including 4K bytes RAM!) it opens many new possibilities for users and systems manufacturers.

You Don't Need an Expensive Teletype.

Using the built-in video terminal and keyboard interface, you avoid all the expense, noise and maintenance associated with a teletype. And the Apple video terminal is six times faster than a teletype, which means more throughput and less waiting. The Apple connects directly to a video monitor (or home TV with an inexpensive RF modulator) and displays 960 easy to read characters in 24 rows of 40 characters per line with automatic scrolling. The video display section contains its own 1K bytes of memory, so all the RAM memory is available for user programs. And the Keyboard Interface lets you use almost any ASCII-encoded keyboard.

The Apple Computer makes it possible for many people with limited budgets to step up to a video terminal as an I/O device for their computer.

No More Switches, No More Lights.

Compared to switches and LED's, a video terminal can display vast amounts of information simultaneously. The Apple video terminal can display the contents of 192 memory locations at once on the screen. And the firmware in PROMS enables you to enter, display and debug programs (all in hex) from the keyboard, rendering a front panel unnecessary. The firmware also allows your programs to print characters on the display, and since you'll be looking at letters and numbers instead of just LED's, the door is open to all kinds of alphanumeric software (i.e., Games and BASIC).

8K Bytes RAM in 16 Chips!

The Apple Computer uses the new 16-pin 4K dynamic memory chips. They are faster and take ¼ the space and power of even the low power 2102's (the memory chip that everyone else uses). That means 8K bytes in sixteen chips. It also means no more 28 amp power supplies.

The system is fully expandable to 65K via an edge connector which carries both the address and data busses, power supplies and all timing signals. All dynamic memory refreshing for both on and off-board memory is done automatically. Also, the Apple Computer can be upgraded to use the 16K chips when they become available. That's 32K bytes on-board RAM in 16 IC's—the equivalent of 256 2102's!

A Little Cassette Board That Works!

Unlike many other cassette boards on the marketplace, ours works every time. It plugs directly into the upright connector on the main board and stands only 2" tall. And since it is very fast (1500 bits per second), you can read or write 4K bytes in about 20 seconds. All timing is done in software, which results in crystal-controlled accuracy and uniformity from unit to unit.

Unlike some other cassette interfaces which require an expensive tape recorder, the Apple Cassette Interface works reliably with almost any audio-grade cassette recorder.

Software:

A tape of **APPLE BASIC** is included free with the Cassette Interface. Apple Basic features immediate error messages and fast execution, and lets you program in a higher level language immediately and without added cost. Also available **now** are a dis-assembler and many games, with many software packages, (including a macro assembler) in the works. And since our philosophy is to provide software for our machines free or at minimal cost, you won't be continually paying for access to this growing software library.

The Apple Computer is in stock at almost all major computer stores. (If your local computer store doesn't carry our products, encourage them or write us direct). **Dealer inquiries invited.**

Byte into an Apple $666.66*

* includes 4K bytes RAM

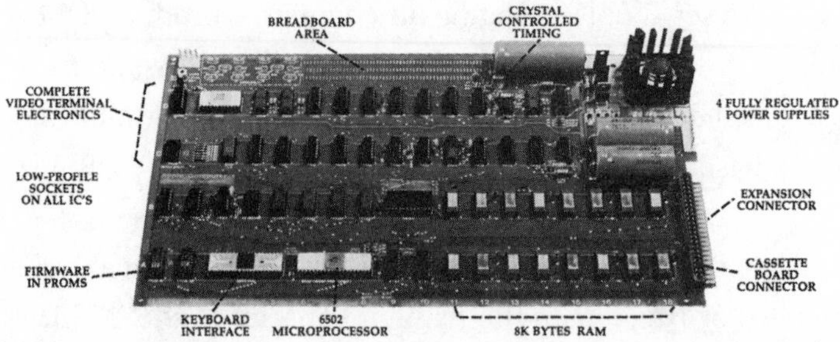

APPLE Computer Company • 770 Welch Rd., Palo Alto, CA 94304 • (415) 326-4248

OCTOBER 1976 CIRCLE NO. 7 ON INQUIRY CARD INTERFACE AGE 11

Apple Computer's advertisement for the Apple 1, "a truly complete microcomputer system on a single PC [polycarbonate] board," published in *Interface Age*, October 1976. Like many first-generation microcomputers of the mid-1970s, the Apple 1 had no monitor, keyboard, or cassette drive for storage, as users were expected to build or source their own peripherals. The Apple 1 also came with no protective casing; what users purchased was only the board. Scanned by Michael Holley for the Wikimedia Commons.

who was still holding down a full-time job at Hewlett-Packard. So while Jobs hustled for sales over summer 1976, Wozniak indulged his inner engineer by exploring what a new iteration of the Apple 1 might look like. For Wozniak, a new Apple was the opportunity to design a computer from scratch rather than as an extension of a prior hobbyist project.[60] By August 1976, Wozniak had a functional prototype that featured several significant advancements over the Apple 1, including color display and a bank of expansion slots, which would allow hobbyists to manufacture their own add-ons.[61]

Yet the immediate path forward was unclear. When Wozniak and Jobs carted the Apple 1 to the Personal Computing '76 Consumer Trade Fair in Atlantic City that August, they left plenty of people impressed with the machine's performance but sold few if any units.[62] Wozniak was uncertain about whether to continue with Jobs, and the two partners disagreed over buyout opportunities. At one point, Wozniak offered the local competitor Processor Technology the option of buying his improved Apple. The pair pitched to Atari, but leadership declined. When representatives from Commodore Business Machines offered to buy the entire company outright, Jobs made an extravagant counteroffer, and a deal never came to terms.[63] Despite the small profit Apple had made from its initial batch of microcomputer sales, it became clear with every passing month that "the rest of the microcomputer industry was growing more quickly than Apple, and Jobs didn't have enough money to match his expanding ambitions."[64] Leveraging his professional networks, Jobs asked his former boss Nolan Bushnell for advice on where to get money. Bushnell advised Jobs that Apple would likely need venture capitalists at some point and suggested his contact Don Valentine, founder of Sequoia Capital and one of Atari's earliest investors.

Valentine was part of a new generation of venture capital investment professionals moving into the consumer technology markets. The history of venture capital is tightly woven into the history of Silicon Valley. Close ties between industry and local universities had spurred technical innovations subsidized by expansive military contracts, while high military demand also catalyzed rapid entrepreneurial activity. The boom in high-tech firms during the 1950s and 1960s made the region a productive place for the formation of investment groups looking to underwrite new business. As Tom Nicholas has documented in his book *VC: An American History*, the region's unique "cluster of economic activity and . . . excess of potential venture-based opportunities" became a self-reinforcing dynamic once venture capitalists began hanging shingles throughout the Santa Clara Valley.[65] Valentine had made early investments in Atari, just a couple of years earlier, which cashed out fourfold when Atari was bought by Warner Communications for $28 million in September 1976.[66] Eager to find further opportunities, Valentine met with Jobs and Wozniak at their garage in Los Altos, but he wasn't convinced. He felt Apple's owners understood neither the larger potential of the consumer computing market nor the demands of marketing (moreover, he thought the greasy-haired, barefoot Jobs was a "renegade from the human race").[67]

But at Jobs's insistence for more contacts, Valentine put him in touch with Mike Markkula, a thirty-three-year-old former tech industry marketing executive who had retired as a "small multimillionaire" thanks to advantageous stock option plays.[68] With degrees in electrical engineering and years spent selling semiconductors for local big boys like Fairchild and Intel, Markkula had a clear understanding of what he was looking at on Wozniak's workbench, and he was "thrilled by the gadgetry."[69] More convinced by Jobs than Valentine had been, Markkula eventually underwrote a bank loan for $250,000 to pay for the development and launch of what was to be known as the Apple II. He also used his professional credibility to lock in the support of the prominent Silicon Valley advertising firm Regis McKenna, which Jobs had desperately been trying to get a deal with.[70] In exchange, Markkula took a third of the company and convinced Wozniak and Jobs to let his old Fairchild chum, Michael Scott, serve as Apple's first president.

Markkula's investment marked a dramatic shift in Apple's capacity for growth, ensuring that Apple could mass produce and mass market its products faster than other homespun microcomputing startups. Markkula's investment was less a show of faith in Wozniak or Jobs or the Apple II and more an illustration of how money creates the conditions for faith to exist. After all, the technology was good, but there was no lack of solid products from more profitable companies; Wozniak and Jobs were both wildly unsuited for management; and the Apple 1 *was hardly selling*—especially not compared to competitors like the Altair, the IMSAI, or Processor Technology's SOL-20. But these were problems that money and management could fix, or so Markkula believed, aiming to rehabilitate Apple's sinking ship through the careful application of marketing, distribution, supply chain, and managerial know-how. Just as Apple's first product had benefited from the general fall in microprocessor prices (thanks to copious support from decades of military-funded research), the region's condensation of successful electronics firms built the first bankrolls for a generation of investors more than aware that new vehicles for capital accumulation needed to be found. Apple, like Atari, would be the beneficiary of this capacious financial hunger.

On January 3, 1977, Apple Computer Company was signed into existence as a corporation and bought out Wozniak and Jobs's simple partnership two months later. Wozniak quit his job at Hewlett-Packard to work on the Apple II full-time, and Jobs was willing, for the moment, to let someone else be in charge. Flush with a quarter of a million dollars, the small group that now comprised Apple—Scott, Markkula, Jobs, Wozniak, and a few friends from high school and Homebrew—set their gaze on an impending target: to have the Apple II ready for unveiling at the first West Coast Computer Faire that April. Like countless technology companies before it, Apple was benefiting from nearly a generation and a half of institutional advantages dealt out to industrial and military technology firms over the course of the mid-century. With Apple, those financial forces were simply finding their latest desire trail in the anticipation of consumer markets.

* * *

By most historical accounts, the first West Coast Computer Faire, held at San Francisco's Civic Auditorium on April 15–17, 1977, was a launch point for the personal computing industry in the United States. Small systems and microcomputer trade shows had happened before, but they tended to cluster on the East Coast. Moreover, they had traditionally been *industry* events, of interest only to hard-core enthusiasts and industry representatives. The West Coast Computer Faire, in contrast, embodied the informal, aspirational counterculture ethos that defined much of hobbyist computing in the greater Bay Area.[71] Organized by Jim Warren and Bob Reiling, both well-known members of the regional computer hobbyist scene, the Faire focused on personal and small-scale computing applications; it aimed to educate and inspire as much as sell and promote. Thirteen thousand people flowed into the Faire over its three-day run, making it at the time the largest convergence of computing enthusiasts in the world, successful beyond anyone's expectations. Markkula, buttonholed on the exhibition floor by *Creative Computing* editor, David Ahl, remarked, "I'm not sure why so many people are here. An awful lot of them are just plain curious as to what's going on. . . . I expected a few more people on a higher knowledge level. I'm very surprised with the whole thing."[72]

Yet even in the midst of the Faire's excitement and chaos, there would've been no missing the Apple II. The company had reserved front-facing booths at the head of the exhibition floor's center aisle, making Apple's newest product the spatial focal point as attendees walked through the Faire's double doors.[73] Whereas most booth holders promoted their wares in front of the Civic Auditorium's standard-issue yellow-gold synthetic curtains, with company names pinned or taped up behind them, the Apple Computer booth was a minimal, modern standout. The few photos that survive the event show intentionally blank, custom-fabricated white walls, only interrupted by a large framed photo of a Red Delicious apple. Installed above is a cleanly manufactured sign showcasing the company's new rainbow-striped "bitten apple" logo, with "Apple Computer" typeset in a chunky, lowercase sans serif—all courtesy of ad firm Regis McKenna.[74] In a market full of boxy machines sold by companies with intentionally high-tech names, Apple Computer exuded a kind of simplistic showmanship. Even Jobs cleaned up for once, tucked into a button-down and slacks with a pinstripe vest and tie. Apple Computer was unrecognizable from the operation that had come to the Personal Computing '76 Consumer Trade Fair just eight months before, where Apple shared a booth with a friendly New York City microcomputer retailer, and Jobs arrived so disheveled that the retailer's mother-in-law demanded he let her stitch up his jeans.[75]

And the Apple II looked the part. Most striking was its light brown plastic chassis, a modestly beveled five-sided wedge encasing a single circuit board with built-in QWERTY keyboard (like many consumer-grade microcomputers of this period, users were expected to supply their own monitor). Acquiring a distinctive case had been one of Jobs's personal missions in the lead-up to the Apple II's launch; his stylistic inspirations included the sleek finish of Hewlett-Packard calculators as well as housewares like home stereos and kitchen appliances.[76] The fact that it was *plastic* was important: plastic casing was expensive but also modern. In a market where most microcomputers were sheet-metal garage constructions, the Apple II's industrial design "bestow[ed] on the computer the identity of an efficient and reliable consumer appliance, not a hobbyist's machine."[77] The design itself came from Jerry Manock, a former product designer at Hewlett-Packard who was barely scraping by as a freelance industrial designer. In Manock's estimation, the design was "conservative," "intended to blend in."[78] Like everything else about Apple's self-presentation that weekend, the plastic case distinguished the company not just from its competitors, but from the hobbyist associations of computing more generally.

Thus the Apple II was more than simply a design object: it was a synthesis of hobbyist instincts for access and expandability woven into a system that was significantly more approachable to novices than anything else available that spring. Inside its casing, the microcomputer was a single board, laid out with Wozniak's typical sensibility for economy of composition. (See pl. 8.) Like the Apple 1, the Apple II used the MOS 6502 microprocessor, which Wozniak still preferred to more popular models, and carried 4K of RAM, though it was expandable up to 48K.[79] Furthermore, eight expansion slots on

A typical configuration for an Apple II when first released in 1977. Note the use of a third-party 9-inch monochrome monitor and an RQ-309DS Panasonic cassette deck, both of which would have been supplied by the user. The Apple II's game paddles, which used a button and a dial for simple computer games, came standard with the system. Image uploaded to Wikimedia Commons by FozzTexx (CC BY-SA 4.0: https://creativecommons.org/licenses/by-sa/4.0/deed.en).

the board allowed users to add specialty peripherals and features to the Apple II. All of this was immediately accessible to the knowledgeable user: the Apple II featured a distinctive plastic "lid" that lifted off without the need to pick up a tool or turn a screw. The ability to open the machine with ease, expand its memory, and add third-party boards indicated Wozniak's sensitivity to what made computing fun for hobbyists: tinkering, exploring, expanding. Making the machine *yours*.

But for users who didn't relish the idea of opening their systems, adding RAM, or learning the finer details of a microprocessor, the Apple II was nonetheless an accessible system. There was no complex process of inputting machine code to get the computer running like the Altair. Instead, a monitor program (one of Wozniak's innovations in the Apple 1) allowed the machine to receive commands on start-up, just like a calculator. Wozniak had also written a version of the BASIC programming language customized for the Apple II and stored on a ROM chip on the board.[80] This meant that users could immediately start programming in BASIC, without taking away from any of the accessible memory available on the computer. The Apple II also adopted the cassette interface used in the Apple 1 for loading and storing software on cassette tape.[81] And because Wozniak loved games, the Apple II was sold with game paddles, capacity for a high-res color graphics mode (including on a standard television set, with the addition of an RF modulator), and even rudimentary, single-tone sound.[82]

Yet despite the Apple II's innovations and prominent location at the Faire, most of the hobbyist press was not sold on it. Apple may have commanded pride of place at the entryway, but its booth was still half the size of its most immediate competitor, Processor Technology, maker of the SOL-20. Jim Warren claimed that he "didn't feel Apple was the strongest competitor," while *BYTE* failed to even mention the company in its coverage of the Faire.[83] Apple took roughly three hundred orders over the next few weeks, surpassing the number of Apple 1s sold in a whole year—a strong showing but by no means a revolution.

By the end of 1977, Apple had new competitors to deal with, namely, RadioShack's TRS-80 and the Commodore PET, both of which also styled themselves as friendly all-in-one microcomputers. Together with the Apple II, these three systems constitute what is retrospectively acknowledged as the "1977 Trinity," the second wave of consumer microcomputers. If the first wave of microcomputers were Altair look-alikes, the 1977 Trinity serves as an index for how quickly the perceived user of the microcomputer was shifting. Gone are the boxy metal cases full of stacked circuit boards, the rows of lights and switches only comprehensible to engineers. Instead, all three microcomputers took the TV terminal as their model, with built-in keyboards and monitors for visual output, as well as cassette storage and onboard BASIC programming functionality. The two new entrants

The Radio Shack TRS-80 (top) and Commodore PET (bottom) microcomputers. Alongside the Apple II, the systems comprised the 1977 Trinity—and were the Apple II's initial competitors. Top image courtesy the Computer History Museum. Bottom image uploaded to Wikimedia Commons by Rama (CC BY-SA 2.0 fr: https://creativecommons.org/licenses/by-sa/2.0/fr/deed.en).

also had operational advantages that Apple lacked: Commodore had been a prominent calculator manufacturer, and RadioShack commanded a chain of electronics storefronts across the nation. These were not hobbyist startups but well-funded corporations with significant capital resources, manufacturing relationships, and retail experience. The TRS-80 proved the most popular system of any of them, at least at first, and both the TRS-80 and the PET were significantly cheaper than the Apple II ($795 and $600, respectively, compared to the Apple II base price of $1,298).

The Apple II, however, turned out to be a long game. While the TRS-80 and PET were less expensive, they lacked the expansiveness of the Apple II's technical features. They weren't designed to be improved on by the user so much as they were designed to be replaced. Both systems lacked the capacity for expanding onboard RAM, meaning consumers were stuck with whatever they had initially bought.[84] The TRS-80 couldn't be opened at all, making it a "hobbyist" computer only in the software sense. In addition, RadioShack's and Commodore's insistence on supplying built-in monitors throttled the immediate expandability of the central board. Apple, in contrast, expected users to supply their own black-and-white or color television—a decision that was perhaps frustrating for consumers who wanted to purchase everything at once but that ultimately allowed for more versatility in output. Also significant in Apple's slow ascent was its early-to-market introduction of floppy disk drives in 1978. Sporting significantly faster load times, floppy disks quickly replaced cassettes as the industry standard storage medium for consumer software. While both of Apple's competitors also released floppy drives, Apple's was far cheaper and did not require the purchase of additional expansion interfaces to be installed onto the board. While the appeal of the cheaper systems played well with consumers circa 1977, Apple's more robust hardware design choices eventually pushed the system beyond its competitors.

Yet beyond the appeal to more novice users, many of the Apple II's features made the system attractive to an emerging class of hardware and software developers. Floppy disk functionality offered a new world of more complex programs versus cassette-based storage. High-res, color graphics were exciting options for budding game developers. And the eight expansion slots (which Jobs initially resisted, wanting a more controlled user experience) made the Apple II the toy of choice for anyone who wanted to hot-rod their hardware. Those expansion slots also made it easier for the Apple II to offer hardware workarounds for some of the system's initial limitations, like its inability to run the popular CP/M operating system or the short character width of its monitor and lack of lowercase letters. Third-party hardware and software developers also found Apple an unobtrusive platform partner. Unlike RadioShack, which initially prevented third-party TRS-80 software from being distributed in its storefronts, Apple made no effort to limit the success of third-party developers.

Over time, this produced a mutually reinforcing dynamic between consumers and developers: as more consumers were attracted to the Apple II because of its breadth of software, developers increasingly focused on that system in order to access the largest potential market—which would prove the source of the Apple II's remarkable longevity. Nothing about what the Apple II might do was preordained at its moment of invention. But what people finally *did* do with it, and how developers tried to accommodate and then to anticipate those desires, is the story that comes next: the story of how the computer became personal.

3

Business

VisiCalc

In the beginning, there wasn't much—software, that is. Flip through old magazines like *MICRO 6502* or *Personal Computing* or *BYTE*, and you get a strong sense of the paltry state of consumer-grade software development circa 1979. There were programming languages for those who wanted to move their Apple II beyond BASIC, rudimentary assemblers and text editors, great swaths of single-purpose applications, and brittle, forgettable games. Software was sold smorgasbord-style, in full-page advertisements unleashing products in long, tumbling lists: *Bridge Challenger*, *Air Raid*, *Diet Planning Package*, *Apple 21* ("black jack for your APPLE!"), *Star Wars/Space Maze*, *Micro-Tax 78* ("just in time to help you prepare your returns"), *Renumber*.[1] Do your taxes, track your calories, save the universe.

The sheer volume and simplicity of this software was an index of how *new* the entire enterprise of consumer software really was. What did people want? For many, it was software that allowed them to rationalize the purchase of a microcomputer to begin with. But what that meant in the late 1970s was still an open question. Ben Rosen, a forty-six-year-old computing industry analyst, described this dilemma in the July 1979 issue of his semiregular publication for Morgan Stanley, *Electronics Letter*:

> Today, virtually the only user of personal computers who is satisfied by the state of the software art is the hobbyist. He does all this programming himself. But for the professional, home computer user, small businessman, and educator, there is precious little software available that is practical, useful, universal, and reliable.[2]

In Rosen's estimation, consumer-grade microcomputers like the Apple II or the TRS-80 had not done much to bring computers to the masses, as many had claimed they would. Whether intended for use at work, home, or school, microcomputing's potential was throttled by a lack of compelling software. Most consumer-grade programs had progressed little beyond simple formulas cloaked beneath a text-based user interface. Imagine dozens of software packages that were simply hyper-specific calculators: one to formulate your gas mileage, one for running statistics algorithms, another for tracking payroll. Users whose needs were more specific got stuck programming themselves. Even Rosen had fallen into that trap, reporting to his readers that he had spent twenty-some hours coding in BASIC just to produce a dividend discount valuation model.

Such was the state of affairs in the microcomputing world—until, that is, *VisiCalc* came along. With *VisiCalc*, the program did not prescribe what was to be calculated, or how. Instead, using paper spreadsheets, or ledger sheets, as a model, *VisiCalc* allowed users to define their content, as well as the mathematical operations they wished to conduct between numerical categories. In other words, *VisiCalc* provided a *framework* for mathematical calculation and modeling that was indifferent to its numerical content. Thus *VisiCalc* offered remarkably flexible uses across financial analysis, engineering, record keeping, budgeting, and other domains—all while ensuring "the user need not know *anything* about computers or programming in order to derive *VisiCalc*'s benefits," as Rosen advised his reader-

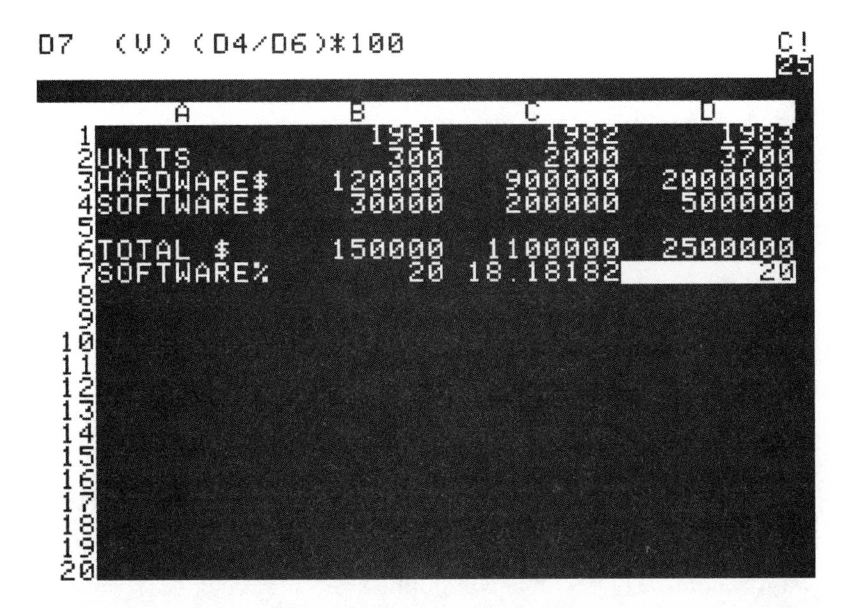

An "electronic worksheet" in *VisiCalc*, better known as a spreadsheet. The screen provides a basic demonstration of *VisiCalc*'s columns and rows. Viewable in the entry line at the top is the equation producing the value in position D7: (D4/D6)*100. Running in emulation on the Internet Archive. Screen capture courtesy the author.

ship of investors, account managers, and consultants.[3] That dividend discount valuation model that took him twenty hours to program in BASIC? Rosen could set it up in *VisiCalc* in just fifteen minutes. Rosen's closing sentiments on *VisiCalc* would become one of the iconic predictions of the early personal computing moment: "*VisiCalc* could someday be the software that wags (and sells) the personal computer dog."[4]

Since Rosen first crowed about *VisiCalc* in 1979, journalists and historians have lauded the program as the "killer app" of personal computing. There is perhaps no piece of software more infamously associated with the Apple II, or the launch of the personal computing industry in the United States, than *VisiCalc*.[5] Developed by two former MIT computer science alums, Dan Bricklin and Bob Frankston, and published by Dan Fylstra, a Harvard MBA graduate, *VisiCalc* was first released for the Apple II in fall 1979. By late 1983, 700,000 copies had been sold across all platforms, many of them at a retail price of $250 (roughly $700 in the early 1980s). These numbers would rocket

Bricklin and Frankston's development company, Software Arts, and Fylstra's publishing outfit, Personal Software, to multimillion-dollar valuations while fueling projections about the profitability of consumer computing. Yet, while *VisiCalc* was a well-conceived piece of software, much of what made it a must-have was the fact that it had no immediate competition. *VisiCalc* arrived at a crux in the consumer microcomputer industry: two years after the 1977 Trinity arrived to market but before any other piece of software had managed to offer a significant value proposition for microcomputer owners who were not programmers.

To understand the success of *VisiCalc*, we must understand the technical, economic, and industrial circumstances that shaped *VisiCalc*'s production, alongside a broader portrait of economic flux that was the context of personal computing's ascent. Looked at narrowly, *VisiCalc*'s development, distribution, and reception help us understand the state of the microcomputing software industry just a couple of years after it made its so-called public appearance. The fact that *VisiCalc* was developed first, and for a couple of years exclusively, for the Apple II highlights important criteria like onboard memory and mass storage as more than just technical minutiae. Viewed with an eye toward industrial relations, *VisiCalc* is also a case study of the emerging developer-publisher relationship in consumer software, which was still formalizing in the late 1970s. The separation of the company that *makes* software from the company that manufactures, markets, and distributes it was a mode of industry relations with few corollaries in the mainframe or time-sharing industries. The particularly contentious dynamics that emerged between *VisiCalc*'s developer, Software Arts, and *VisiCalc*'s publisher, Personal Software/VisiCorp, showcase the economic interdependencies and differential risk developers and publishers negotiated as a burgeoning microcomputer software industry rapidly scaled during the early to mid-1980s.

But beyond the specifics of *VisiCalc*'s production, this program also provides a necessary introduction to one of microcomputing's most broad-ranging uses: business applications. The months and years following *VisiCalc*'s release would bring a swath of business-oriented programs to the Apple II (and other microcomputer systems), including spreadsheet and other kinds of database software, inventory management systems, invoicing packages, tax assistance software, and a host of industry-specific software suites across occupations as distinct as farming, dentistry, and construction. *VisiCalc*'s popularity—and that of the microcomputer hardware and software industries with it—was first driven by an expanding base of non-hobbyist white-collar professionals who reveled in deploying the program for projection, forecasting, modeling, and other modes of professional management. Under the hood of *VisiCalc* is a transformation in the technics of work that has become so commonplace that today it is hardly noticed: the way "computerizing" one's business became an aesthetic experience unto itself, rife with sensations of transparency, immediacy, and control.

Arriving at the sunset of the 1970s, *VisiCalc*'s popularity among the Wall Street set was a modest historical affair compared to the larger geopolitical issues of the day. By 1979, the United States sagged under spiraling inflation, rising unemployment, ballooning deficit, and a torpid global energy crisis that President Jimmy Carter was failing to redress through energy conservation legislation.[6] Faith in American government had reached unprecedented lows. By the middle of the year, Carter addressed the nation on its "crisis of confidence," the "growing doubt about the meaning of our own lives and . . . the loss of a unity of purpose for our nation."[7] Carter's words gave language to a palpable sensation that American prospects were beset by uncertainty.

Yet while some mourned the loss of the traditional symbols of American power—the factory, the steelworker, the assembly line—others would double down on free market ideologies that promoted a neoliberal political economy grounded in "liberating individual entrepreneurial freedoms," in the words of the sociologist David Harvey.[8] This was a purpose for which *VisiCalc* was uniquely suited. As we shall see, themes of the entrepreneurial individual course through *VisiCalc*'s advertising and journalistic coverage, as do anxious references to time, energy, economic surveillance, and the internalization of personal responsibility. *VisiCalc*'s affordances were presented not just as a handy helpmate for business, but as a tool for bracing oneself against uncertainty: "With *VisiCalc*, getting your arms around the future seems a trifle easier."[9]

* * *

Dan Bricklin didn't dream of making software for personal computers, bringing computers to the masses, or disrupting oppressive institutional hierarchies through the power of computing. Like most computer entrepreneurs of his generation, his ambitions were simple: make software, make money. And his pathway into computing was about as predictable as they come. Growing up in Philadelphia in the 1950s and 1960s, Bricklin took to electronics hobbyism like many young white men of his generation who showed a proclivity for math and science. Through family connections, he was allowed to tinker on a local school's mainframe, and the nearby University of Pennsylvania granted him the opportunity to teach FORTRAN to Wharton graduate students while he was still in high school.[10] Sticking to the East Coast, Bricklin received a bachelor's degree in computer science from MIT in 1973, followed by several years' experience working at Digital Equipment Corporation (DEC), the popular minicomputer manufacturer headquartered just a few miles west of Boston.[11] When he returned to school in 1977, it would be to climb the trellis of the Ivy League, enrolling in the Harvard Business School (HBS) because of concern that computer programming lacked potential for long-term financial security.[12] *VisiCalc* would originate where upwardly mobile white men were refashioning themselves for an economy that was leaving many others behind.

At HBS, Bricklin found a world of people who could greatly benefit from computer applications for instantaneous modeling and financial synthesis, and especially from software that didn't require much in the way of programming chops. The evidence of this unmet need was everywhere: from the tedious process of writing his own financial projection programs on the university's PDP-10 time-sharing system to "running the numbers" by hand for his homework case studies to observing his professors, clouded in chalk dust, onerously recalculate financial tables every time they adjusted a model on the blackboard.[13] As he sat in class, Bricklin's mind wandered. He imagined a heads-up display "like in a fighter plane," a trackball in the back of a calculator, a world of command-and-control hardware interfaces bestowing godlike perspective on the world's numbers. The tech journalist Robert X. Cringely would gloriously embellish the scene as one in which Bricklin, "like Luke Skywalker jumping into the turret of the Millennium Falcon, . . . saw himself blasting out financials, locking on the profit and loss numbers that would appear suspended in space before him"—an unironic reading that only exacerbates the propulsive machismo of it all.

These were, of course, hardware fever dreams. Nothing like this was plausible, not within Bricklin's reach anyway. So he scoped his fantasies downward over summer 1978, into the realm of material feasibility. He imagined the general schema of what would be called an "electronic worksheet" in the future program's as-of-yet unwritten documentation, later simply known as the *spreadsheet*: a digital, interactive, real-time mathematical matrix that collapsed the typically distinct computational processes of input, calculation, and output into a seamless user experience. His plan for profitability, however, was more in keeping with the industrial spirit of the Death Star than the wily guerrilla tactics of the Rebellion: he planned to design his program for a DEC time-sharing system, then hawk it to businesses and institutions. He knew little of the hobbyist computer world, had no visions of hackers admiring his stylish code. *VisiCalc* was imagined as what we would today call *enterprise software*: just the logical extension of a market opportunity.

Now, enterprise software wasn't a new concept; computers had been deployed for business and administrative purposes since the early 1950s.[14] By the late 1970s, a software products industry for mainframe and minicomputers was fairly well established, and nearly all of this product would have been oriented to use in business or industrial settings.[15] This included a wide range of systems and applications software, ranging from operating systems to programming aids and utilities to industry-specific programs for banking, engineering, insurance, and transportation to cross-industry applications for handling inventory management, payroll, marketing, and general accounting. By 1980, the cumulative revenue of the US packaged software business reached nearly $2.5 billion.[16] It is unclear to what extent Bricklin thought through the challenge of developing new software in such an established and consolidated market (where the top 15 percent of all software suppliers took 68 percent of all revenue), but with his Ivy League credentials and unique blend of computational and financial know-how, his schemes may not have seemed entirely outside the realm of possibility.[17]

More intriguing than Bricklin's aspiration, however, is the way his educational and professional background shaped how he thought about the relationship between hardware, software, and the people on the receiving end of these technologies: the users. As a student at MIT in the early 1970s, Bricklin had the advantage of working on the cutting edge of new and innovative time-sharing systems—systems that, if not technically "real-time," afforded an experience of interactivity and immediacy by virtue of how time-sharing worked. In his post-MIT work at DEC, Bricklin was assigned to the company's Typeset-10 word processing system, which allowed users to prepare raw text in a format appropriate for a typesetting machine (such systems were increasingly common at newspaper offices, among other places: part of a larger trend in the computerization of labor). Specifically, Bricklin was tasked with the software development for the Typeset-10's editing terminals, which worked in real time; users had to see what they were writing as they wrote it and be able to correct errors as they went.[18] Features he worked with on the Typeset-10, like an embedded on-screen ruler, keystroke minimization, and on-screen scrolling, focused on end-user efficiency and ease of use. Thus the functionality Bricklin envisioned for his future software product expressed a user-centric philosophy. He designed from the point of view of a working professional, *not* a hobbyist, and he could think in those terms because, thanks to his experience working with time-sharing systems, he expressly did not imagine himself operating within the limitations of a microcomputer.[19]

The impetus to look toward microcomputers came via the privileges of his Harvard network. In fall 1978, Bricklin (now accompanied by an old MIT friend turned development collaborator, Bob Frankston) was directed by a Harvard finance professor to get in touch with another recent HBS graduate, Dan Fylstra (also an MIT engineering graduate).[20] Fylstra was notable among recent HBS graduates for his fervent faith in the future of personal computing, having founded a microcomputer software mail-order business while finishing his second year at Harvard.[21] As computing entrepreneurs on the East Coast went, Fylstra was one of the best positioned to understand the undercurrents of the microcomputer market: he had been a founding associate editor of *BYTE* in 1975, writing the magazine's first reviews of the Commodore PET and the TRS-80.[22] It was the kind of institutional coziness that helped his mail-order racket, Personal Software, grow into one of the nation's most successful microcomputer software publishing businesses by the time he was in conversation with Bricklin and Frankston barely eighteen months later.

Fylstra was adamant that Bricklin should develop the program for the microcomputer rather than a DEC time-sharing minicomputer. This advice was one part prescience, one part self-interest: software sold directly to corporations went through entirely different distribution channels than consumer software, and Fylstra's business was not structured to support the former. In order to get a cut, Fylstra needed Bricklin to develop something he could sell through the industrial pathways he already held sway in: the emerging microcomputer industry.

The decision to pivot *VisiCalc* to microcomputers was a significant change in Bricklin's mental map in terms of both programming design and intended market—though, as far as the archival record suggests, it prompted no particularly dramatic concern. "Saw Fylstra and got Apple manuals," Bricklin wrote in his production diary, summarizing the events of September 25, 1978.[23] Bricklin's willingness to develop for a microcomputer rather than a time-sharing system wasn't based in a passion for the hack or a desire to prove anything about the potential of microcomputing at all but in a deliberate set of business decisions largely governed by his most interested potential investor.

Over the next two weeks, Bricklin learned what he could about the Apple II, then crunched out his first prototype in BASIC over the first week in October on Fylstra's Apple II. *VisiCalc*'s initial prototype was a rudimentary thing, lacking many of the advanced features Bricklin hoped for, but Fylstra liked what he saw and conversations continued.[24] In Bricklin's notes, he called his prototype "FINANCE TEST PROGRAM" (the Apple II only wrote in uppercase); other sources refer to it as an "electronic blackboard," and Bricklin even wrote a paper for his marketing class about the product, which he called "Calcu-ledger."[25] The name *VisiCalc*, short for *visible calculator*, would come later.[26] What Bricklin had written in BASIC was simple, but the foundations of *VisiCalc*'s future were there: columns and rows and the ability to change calculations with a keystroke—an entire responsive, interactive mathematical world, glowing on screen.[27]

* * *

From a business perspective, Fylstra's decision to point Bricklin toward the Apple II may have seemed incongruous. Of the three systems making up the 1977 Trinity, the Apple II had sold the least, and not by a small margin. By some estimates, Commodore's PET was outselling the Apple six to one, and the TRS-80, with its commanding position in the RadioShack distribution network, was outstripping even that.[28] The Apple II was the most expensive and yet least outfitted of the three platforms, making it no surprise that many early adopters contented themselves with less expensive systems. As indelible as *VisiCalc*'s association with the Apple II would become, it's unclear to what extent the decision to develop for the system was ever a topic of explicit debate or consideration between the three men. Retrospectively, Fylstra has stated that his preference for the Apple II grew from a number of factors, including his appraisal of the system's technical design, Apple's sophisticated marketing, and Personal Software's own sales data, which indicated growing strength in Apple software sales.[29] Given his industry position and his early experiences at *BYTE*, it's clear that Fylstra had exposure to all three systems and a strong sense of their comparative advantages and limitations.

Thus, despite its smaller installed base, the Apple II sported a matrix of technological and industrial qualities that, when laid atop one another, gave the platform significant affordances over its competitors. These included its high memory threshold, its early-to-market floppy disk mass storage devices, and Apple Computer's hands-off attitude to developers. Each of these qualities contributed to the Apple II's unique suitability for the technological requirements and economic goals of *VisiCalc*.

Memory is essentially a measure of how much "room" a computer has for holding instructions, which is the basis of what allows computer hardware to run software; software must be "loaded" into memory in order to be executed and manipulated. In 1975, the least expensive Altair came with 1K of memory, though it required at least 4K to run *Altair BASIC*. What Bricklin and Frankston proposed just a few years later was far more memory-intensive: they aspired to a minimum of 16K when they started prototyping, though the program ultimately required 32K.[30] Yet of the 1977 Trinity, only the Apple II had immediate capacity to meet such a high memory threshold. The initial 1977 releases of the TRS-80 and the Commodore PET both topped out at 8K. The TRS-80 could go up to 48K by 1978 but required the additional purchase of an interface extension module. The PET would not get a technical redesign until 1979.[31]

In contrast, the Apple II, from inception, was guided by Steve Wozniak's homebrew orientation to do-it-yourself expandability. It could take upward of 48K at the moment of release, and expanding memory or adding peripherals to the Apple II required no interaction with the company or further commitment to proprietary hardware (unlike the TRS-80). These sorts of details would have been nuts-and-bolts knowledge for anyone paying attention to the microcomputer hardware industry, making it entirely possible that the Apple II was not even so much a "choice" as it was the only reasonable option for development. While a fully outfitted Apple II was extraordinarily expensive in spring 1977—$2,638 at the time, or about $11,000 in 2020—memory costs predictably went down rapidly over the next few years, making Fylstra's bet on the Apple II less risky in the long run than it might have seemed in the immediate.

Apple Computer's release of a floppy disk mass storage peripheral, the Disk II, in summer 1978 contributed further to the Apple II's advantages as a development environment for resource-intensive software like *VisiCalc*.[32] While we often think of mass storage devices like USB drives or even cloud computing as a way of making data transportable, mass storage devices of the 1970s were more about providing a way to "mak[e] up for the limitations of a finite memory in any particular computer's implementation," as Carl Helmers wrote in a spring 1976 issue of *BYTE* devoted to the subject of cassette storage.[33] First- and second-generation microcomputers had no hard drives as we are familiar with today. Mass storage thus allowed users to maintain a library of programs and files that were external to the computer itself. This was a crucial convenience during a time when the only other way to "load" software into a computer like the Altair was to tediously key data instructions into the computer's random-access memory using the switches on the machine's front panel (or, later, to type a program into memory in BASIC).[34]

Of the various mass storage options available to hobbyists and other consumers in the mid-1970s—paper tape, cassette tape, and floppy disks—cassettes had quickly dominated as the mass storage medium of choice.[35] While cassette tapes are more widely remembered as a vintage form of *audio* media, binary data could be "inscribed" on a cassette's magnetic tape as cycles of analog waves, interpretable by a microcomputer's cassette interface adapter.[36] The popularity of this format was driven by the low cost and ready availability of both tapes and recording decks, forging an assumed standard visible in the hardware design of many late 1970s microcomputer systems. The

Close-up of Apple's Disk II 5.25-inch floppy drive peripheral. Floppy disks were inserted horizontally into the front-facing slot and secured by pushing down the black flip-up tab to close the slot. To read the disk, a moving element called the "read/write head" would scan the disk's magnetic field and interpret those patterns into an electrical current the Apple II could process; this activity was reversed to inscribe data onto the disk. The adapter on the side was connected to a Disk II expansion board plugged into the Apple II motherboard. Image courtesy Andrew Borman.

Commodore PET and the TRS-80 both came with cassette decks as part of the purchase price.[37] In contrast, the Apple II only offered the cassette interface hardware, not the deck itself; ever the hobbyist, Wozniak presumed users would want to customize by making such decisions for themselves.[38]

Despite their low cost and accessibility, however, cassette tapes had disadvantages. Because the magnetic substrate of a cassette tape is essentially one long ribbon, data must be stored sequentially. Thus a cassette tape could only hold information bit by bit; when reading a program on cassette tape, each bit had to be individually loaded into the computer's memory. In practice, this resulted in very slow load times for cassette-based software. Moreover, cassette indexing was not automated: the cassette itself did not "know" where different pieces of data were on the tape. This resulted in a demanding and time-consuming process of searching for one's data, as reported by one early microcomputing consumer guide:

> Suppose, for example, that you are the secretary of the local computing club and thus have the job of maintaining the mailing list which is stored on cassette. You will quickly discover that changing an address or adding a new name is a somewhat time-consuming process with tape. How long would it take to arrive at a destination in downtown Chicago if the cab driver starts at O'Hare and must go by every house between the airport and downtown? That, essentially, is what happens with tape storage.[39]

By some estimates, a user could spend thirty to sixty minutes trying to locate data on tape.[40] While cassettes may have been a serviceable distribution format for some software programs, these data access constraints made cassettes a poor format for a product like *VisiCalc* that was designed for users to create many saved files of their work.[41]

Floppy disks resolved the limitations of cassette tapes by allowing for a more sophisticated information architecture across the magnetic surface of the disk. Invented by IBM in the early 1970s for their low-end System 3 mainframes, floppy disks were quickly adapted by several companies into mass storage technology more appropriate for corporate desktop data processing systems and, later, consumer-grade microcomputers.[42] Rather than a long strand of ribbon, the essential component of a floppy disk is a thin circle of magnetized film, stored in a flexible, protective plastic sheath (giving floppy disks their distinctive floppy "wobble"). The surface of a floppy disk's film is laid out in concentric tracks crosscut like pie into sectors, all invisible to the human eye.[43] This allows every set of data on a floppy disk to have a known location, guided by an index. To continue the taxi driver metaphor, the difference between a cassette and a floppy disk is that while a cassette stops at every door, a floppy disk holds a map containing knowledge of where data are located on the disk. While floppy disks couldn't hold as much data as a cassette tape, their method of data

Image of a 5.25-inch floppy disk (top) and a diagram illustrating the disk's data architecture of sectors and tracks (bottom). The floppy disk's protective jacket enclosed a circle of thin magnetized film. When it is inserted into a drive, a motor spins the film inside its jacket, allowing the data to be read through the oval-shaped area of exposed film at the bottom of the disk. Photo courtesy Tega Brain, diagram by the author.

organization permitted drives to find and read data far more quickly than a cassette deck could. As one user reported in 1980, "Despite a higher initial investment, the floppy disk is more reliable, and it can transfer programs and data as much as 30 times faster than the audio cassette."[44] In the world of software industry trade-offs, a more efficient user experience won out over sheer storage capacity. While cassette tapes remained popular in hobbyist circles, they would be quickly abandoned as a commercial distribution format once floppy disk drives became more affordable.

It would take several years, however, for the price of disk drives to reach the threshold of consumer affordability. Eight-inch drives were common enough in corporate environments, but desire for a lower-cost drive inspired the creation of the 5.25-inch disk, initially known as a "mini-disk."[45] By summer 1978, both RadioShack and Apple had released floppy drives for their systems, allowing users to upgrade beyond cassette storage—though while Apple's drive simply plugged into one of the system's eight expansion slots, the TRS-80 required the purchase of the system's additional expansion interface unit, making the floppy drive a peripheral of a peripheral.[46] While the drives Apple used were third-party appliances, Wozniak redesigned the controller card to use 75 percent fewer chips than their original equipment manufacturer's off-the-shelf card did, helping boost the disk drive's affordability and profitability at the price of $495 and up.[47] While the initial Disk II sold for more than three times the cost of its parts, it was still the least expensive disk drive on the market. As Frank Rose claimed in his late 1980s account of Apple's history, *West of Eden*, the Disk II "suddenly transformed the Apple II from a gadget only hardcore hobbyists would want to something all sorts of people could use. . . . In strategic terms, Woz's invention of the disk controller was as important to the company as his invention of the computer itself."[48]

The Apple II's final advantage came not from technical capacities but from how the hardware company created a constructive environment for independent software development. First, Apple had made a serviceable effort to document the platform's source code, giving aspiring software developers enough information on the Apple II's internal workings to effectively develop code for the platform. These decisions helped propel development on the Apple II, especially for software like *VisiCalc* that was programmatically complex. Furthermore, in hobbyist fashion, Apple placed no restrictions on developers who wanted to design and market software for the Apple II.

While it may seem sensible in retrospect that more software would enhance a platform's value in a competitive marketplace, this was not the approach taken by the microcomputer industry's earliest leader, RadioShack. RadioShack sought to command the TRS-80's software offerings at the level of retail distribution, essentially forbidding RadioShack franchises from selling or even mentioning software that was not published by RadioShack itself. The early popularity of the platform had driven many developers to write their first software for the TRS-80, but they soon discovered it was tremendously challenging to gain visibility for their software since they couldn't get a point-of-sale presence in stores.[49] As an avid surveyor of not just these microcomputers' technical affordances, but their distributor and retail relations, Fylstra would have been fully aware that releasing *VisiCalc* for the TRS-80 would have required giving RadioShack a cut of his revenue.

Altogether, the advantages that made the Apple II an ideal development environment for *VisiCalc* were not so much the platform in isolation as the way it operated within a competitive ecosystem: it had a specific set of affordances that suited both the technical and economic desires of its creators. For technical reasons, the Commodore PET was never a consideration. And while a souped-up TRS-80 could have met the minimum memory requirements for software like *VisiCalc*, RadioShack's policies on software development and distribution discouraged ambitious and innovative third-party development for the platform. The historic linking of *VisiCalc*, the Apple II, and the rise of personal computing reveals itself as a set of nested interdependencies between economic, technical, and practical concerns about what made computing "easy" to the most minimally trained user.

* * *

Whether Apple II users made do with their cassette deck or upgraded to the newest mass storage device to run *VisiCalc*, they were nonetheless operating in a new world of software production: software sold in individual units to individual consumers and privately owned, the way records, 8-tracks, or VHS tapes were. This was a distinct transition from mainframe or minicomputer systems, where software was typically owned or leased by the company that purchased it or developed it, never by any single user. Both cassettes and floppies offered an inexpensive, easily reproducible format for commercial software distribution, and as they lacked proprietary format restrictions, software programmers could duplicate their software on cassette or disk themselves at no cost beyond time, labor, and material goods—resulting in extremely low barriers to market entry.

Making the sale of software *profitable*, however, was a different proposition. This required an understanding of how to efficiently manufacture, distribute, and market software—a wholly different set of skills from simply developing code. Such particularities are already evidenced in the story of *VisiCalc*: would the program have existed at all, or achieved the overwhelming impact it did, without the intervention of Dan Fylstra and his insistence on developing it for microcomputers? Fylstra's ability to influence such commercialization decisions, despite not creating the software itself, is just one illustration of the complex relationship between software *developers* and software *publishers*, which emerged unevenly over the course of the mid- to late 1970s.

The earliest formations of the microcomputer software industry were sometimes compared to the more familiar relationships between record labels and musicians or publishers and writers.[50] Both of these models separate the creative and material labor that creates a unique cultural product from the organizational and social labor that moves that product through modes of capitalist exchange. Such an arrangement was often believed to allow each party's talents to thrive by not putting undue responsibilities on either. Like most musicians and writers, software developers typically did not have the institutional know-how, social relationships, or startup capital necessary to engage in the costly process of self-managed manufacturing, distribution, and promotion of their programs. By tying themselves to a publisher, they gained commercial expertise and resources. Meanwhile, software houses usually did not have the organizational temperament to internally develop the volume of content they needed to achieve profitability at scale.

When there was barely a software industry, of course, there were no independent publishers acting as middlemen between developers and consumers, because there was no room for them in the ecosystem. Remember that Micro-Soft's *Altair BASIC* was sold directly by MITS and that hobbyists readily circulated their own software for free within their communities (to say nothing of the prevalence of software piracy, especially among hobbyist groups).[51] Microcomputer software developers interested in designing operating systems and programming languages generally approached, or were approached by, hardware manufacturers, bypassing the commercial marketplace. But consumer applications changed all that. Applications software—whether games, word processors, or a program like *VisiCalc*—competed side by side in magazine ads and on retailer shelves for command of consumers' attention. In many cases, especially from 1975 to early 1977, these products were simply sold by the same people who programmed them, as attested by the quaint, unpolished style of many early software advertisements.

For microcomputer enthusiasts with a dash of financial know-how and marketing expertise, however (say, for example, an HBS graduate), the market opportunity would have seemed obvious. Manufacturing software on floppy disk was fairly inexpensive, but it could always be made cheaper if more units could be duplicated or blank disks ordered. Advertising space in computer enthusiast magazines was relatively inexpensive in the mid-1970s, but the more products someone could market, the less they might pay on columns per inch in print. And the more products sold, the more leverage one had with distributors, retailers, and journalists. Individual developers couldn't do much to make software more quickly to achieve these kinds of advantages, but a savvy communicator and confident negotiator could convince other developers to let them do the heavy lifting of manufacturing, marketing, and distribution for them—all while skimming the advantages of such economies of scale.

Such was the path of Dan Fylstra and the publishing outfit he began in 1977, Personal Software. The business started as nothing more than a mail-order software operation run out of the apartment he shared with his wife, Hillary, in Cambridge, Massachusetts, where they lived up to their elbows in cassette tapes, documentation sheets, and packing materials.[52] Extraordinarily little is documented on the earliest history of Personal Software, though some details can be reverse engineered by studying early advertisements. Personal Software's first product was likely Fylstra's own creation, a 6502 assembler for the Commodore PET.[53] Beyond this, the company's earliest ads were a smattering of whatever Fylstra could get his hands on, mostly for the PET and the TRS-80. He had a habit of bundling separate programs, often by separate developers, into cheap "packages": a graphics package comprising DOODLER, PLOTTER, and LETTER; a "personal finance" suite made of INSURE, HOME, and SPEND; even a collection of ten simulation games for the bogglingly low price of $14.95.[54]

This business model relied on establishing relationships with numerous developers. The economic arrangement that typically governed such relationships was the royalty structure, in which the developer was given a percentage of the revenue taken on each sale, based on invoices over a given period, often monthly. Exact percentages for Personal Software's earliest products are unknown, but in the industry overall, royalties varied anywhere from 5 to 25 percent of each unit a publisher sold and were often accompanied by an advance that helped fund a programmer while they developed and polished their software.[55] These royalty percentages applied to both wholesale revenue (which was the price distributors paid before marking up the product and sending it down the supply chain to retailers) and the list prices paid on any mail-order or direct sales handled by the publisher itself.[56]

Under a royalty structure, developers were contractors, not employees—an understandable arrangement at a time when "microcomputer software designer" was not yet a job description and it was unclear how sustainable such work would be in the long term. The stress of maintaining stable cash flow undoubtedly also made programmer royalties an appealing compensation structure for emerging publishers, allowing an undercapitalized company to expand its software library while mitigating the risk of increasing operating expenses from payroll—ensuring the developers only got paid when publishers got paid.

By late spring 1978, Personal Software was no more noteworthy than any number of quasi-publishers advertising in popular computer hobbyist magazines.[57] Sometime that year, however, Fylstra made the acquaintance of Peter R. Jennings, a Canadian developer in his midtwenties who had programmed a popular chess game, *Microchess*, for the single-board KIM-1 microcomputer.[58] Fylstra first struck a deal to become Jennings's publisher and then, later, brought him on as a partner at Personal Software. After converting the software to function on the TRS-80, the Commodore PET, and the Apple II by 1979, Personal Software was able to laud *Microchess* as the industry's first "Gold Cassette," selling more than fifty thousand copies.[59] Yet while games made up nearly half of Personal Software's offerings by fall 1978, Fylstra desired to move into "more serious software": code for professional and business applications.[60] Where games were discretionary, business software could become infrastructure—and could be priced accordingly, despite being manufactured and advertised at the same cost to the publisher.

This is precisely the market opportunity Fylstra sensed while discussing *VisiCalc* with Bricklin—an opportunity significant enough that he initially tried to convince Bricklin to become an employee of

Personal Software, operating inside his own business.[61] While the archival record is not explicit about Fylstra's motivations, the economic potential would have been apparent given his understanding of the scope and scale of *VisiCalc*'s potential market. For publishers, a primary advantage of royalty deals is that they form a kind of insurance for recouping expenses, especially for a product that might underperform, insofar as the publisher is prioritized by the percentages to collect the majority of revenue. But all royalty arrangements have tipping points, a threshold where a product sells so well that royalties become perceived as revenue *lost* to a publisher (given that a publisher might have captured more revenue by paying its developers as employees). Notably, Bricklin declined the offer to work for Personal Software (he had, after all, enrolled at HBS to get out of the trap of simply being someone else's programmer). Bricklin's desire for autonomy as a software developer, despite the additional bureaucratic overhead, would have a long-term impact on *VisiCalc*'s future. The program's success, and later waning, would prove less about the software itself and more about the commercial architectures it was wrapped up in.

No formal agreement governed the early talks between Bricklin, Frankston, and Fylstra. Fylstra's support was at first casual and loosely defined. He provided Bricklin with information on the Apple II, and eventually with the machine itself, and engaged in iterative feedback with Bricklin and Frankston as they developed the idea in fall 1978.[62] The unstated commitment between the parties was common enough in this very early stage of industry, where outcomes and markets were still quite formative. Nonetheless, a verbal royalty arrangement was set early in the development process, before any formal publishing agreement between the two entities, at the rate of 35.7 percent. No industry best practices existed yet to govern what a royalty arrangement should look like, but Fylstra and Jennings have admitted it was high for the time, though not unheard of.[63] A formal publishing arrangement emerged in spring 1979, sometime after Bricklin and Frankston formalized themselves as a development studio called Software Arts that January.[64] In addition to laying out royalty stipulations, the contract established both companies' interdependencies: Personal Software was restricted from developing or marketing any other spreadsheet product, while Software Arts had to follow Personal Software's marketing direction.[65]

While Bricklin and Frankston got to work, Fylstra and Jennings's job, as their publisher, was to prepare the runway for *VisiCalc*'s arrival. This involved a variety of tasks that reflected the range of support early publishers provided for their clients, including industry outreach, manufacturing, distribution, and—perhaps most important—provision of startup capital. Personal Software sank a reported $100,000 of *Microchess* revenue into *VisiCalc*.[66] Some of this money constituted an advance royalty payment to Software Arts to support development costs, including the monthly lease of a timesharing system.[67] Jennings even elected to forgo his royalty payments on *Microchess* for a while to give Personal Software more cash flow while *VisiCalc* was being developed—an indication of just how undercapitalized even the most successful early publishers were.

Aside from supporting early development costs, Fylstra and Jennings arranged to demonstrate early versions of the program to key industry players, including Mike Markkula, Steve Wozniak, Ben Rosen, and Fylstra's old *BYTE* chum, Carl Helmers. In every case, the goal was to prime interest and solicit feedback, especially with regard to marketing and pricing.[68] Lacking a model for the release of this kind of software, conversations with industry leaders helped Fylstra and Jennings maximize potential profit by adjusting their price to what they thought the market would bear. They had initially imagined a price of around $35—more than a game but less than many programming languages or operating systems—but they marked up the price to $99.50 on launch.[69]

Large-scale conventions, like the National Computing Conference and the West Coast Computer Faire, served as natural anchors for many of these meetings, as well as an opportunity to run sneak peeks with the press. The pair also met with representatives from Apple's marketing department, who helped give Personal Software an edge in understanding the intricacies of Apple's distribution network. As Fylstra recalls:

> With help from Apple, I was able to give presentations of VisiCalc at several regional distributors' meetings, with hundreds of computer store managers in attendance. . . . The effect of scrolling the small Apple display to reveal a seemingly infinite worksheet, and of changing one number and seeing all the calculated values change, was electric on the audience. Computer store owners and their salespeople soon realized that they could sell an entire Apple Computer system and VisiCalc software.[70]

This was all part of the emerging supply chain that publishers handled, in which their role was to educate distributors, and the retailers who purchased from them, on the operation and benefits of a particular software package. While retailers were largely in control of what product they sold in their stores (provided their distributor carried it), they preferred to push software from publishers that they felt they had a relationship with or that provided retailer training. The depth of investment required to support these kinds of interactions with distributors and retailers showcases the fundamental division of labor between the developer and the publisher. Such hands-on activities were well beyond the scope of what Bricklin and Frankston could have achieved while working on *VisiCalc*. In order to be closer to this kind of institutional knowledge, Fylstra and Jennings also made the decision to move Personal Software to California in May 1979. This was a significant transition, insofar as it left their primary partner, Software Arts, on the East Coast—a geographic separation that punctuated the distinct roles the two companies were playing.

* * *

In New York City, somewhere near the corner of Madison and 30th—far from Personal Software or Apple Computer or any of the men who saw themselves as the industry's kingmakers—Stan Veit was watching something happen. Businessmen kept coming into his Midtown store, Computer Mart, and asking to buy a "VisiCalc machine."[71] It wasn't something that happened all at once but over a tight span of years, as 1979 ran into 1980, 1981. In their wide ties and three-piece suits, they didn't care about programming for programming's sake, or learning ROM from RAM. Veit was watching computer habits coagulate in real time: "They [customers] were more interested in what a computer did than in how it did it."[72]

VisiCalc may have provided a rationale for "buying a ten-thousand-dollar computer system to run a one-hundred-dollar program," but *who* was in a position to buy either platform or program points to a marked departure from the traditional microcomputer enthusiast—as Veit's firsthand experiences attest.[73] Many of *VisiCalc*'s earliest customers were wealthy, white-collar men interested in microcomputing as a tool to amplify their professional lives. Some would catch the computer bug, yes, but most experienced their newfound computer hobbyism (if they identified as hobbyists at all) through preprogrammed software applications. This was so-called transformative technology bent to the least transformative of goals: getting ahead of your competition.

Fylstra and Jennings had played to these business users' anxieties and desires with *VisiCalc*'s manicured marketing strategy. In an era when most software was sold in Ziploc plastic bags with one sheet of photocopied documentation, Personal Software handled *VisiCalc* as a "whole product," a marketing concept Fylstra claims to have learned at HBS.[74] The premise was to treat the experiences of handling and of learning to use *VisiCalc* as aspects of the product and to make this "whole product" easily understandable to nonprogrammers, at every level of interaction. This meant extensive documentation, reference guides, model spreadsheets, and packaging that made users feel like they were dealing with a serious product—all of which was a first in the microcomputer software industry. *VisiCalc* came packaged in a brown vinyl folder, reminiscent of a leather attaché case or executive binder (see pl. 9), along with a hundred-page manual divided into four lessons and a quick reference card.[75] Fylstra especially had a sense for the cultural work necessary to incorporate the Apple II's unfamiliar technology and *VisiCalc*'s nuanced functionality into the professional imagination; everything about *VisiCalc* was promoted as easy, helpful, professional, and progressive.

This sensibility was potently expressed in *VisiCalc*'s first advertisement, produced for Personal Software by Regis McKenna, the same public relations firm Apple used (and a relationship brokered for Fylstra by Steve Jobs).[76] It was the first full-page, four-color-process ad for a microcomputer software product ever printed, a slick and shiny thing telegraphing a new world of expectations for what personal computing software could become (see pl. 10).[77] It is also a case study on the anxieties of American business practice circa 1979. "Solve Your Personal Energy Crisis," announces the headline, a not at all wry nod to the ongoing 1979 Oil Shock. Unfolding in the wake of the Iranian Revolution, the Oil Shock was the second energy crisis of that decade, catalyzing panic buying, gas rationing, and generalized economic anxiety throughout the United States (Jennings himself recalls having to take the train from Palo Alto to San Francisco for the 1979 West Coast Computer Faire because of the gasoline shortages).[78] This wasn't software for some kind of countercultural utopia but software that lived in what was real and immediate, even if it was targeted only to the class of people already most privileged and able to weather the economic storm—in other words, the white male professional, so prominently featured in the ad itself.

VisiCalc's first ad collapses the personal and the geopolitical, scaling the impact of widespread infrastructural precarity to the level of one man's frantic calculations. In the advertisement's scene, our everyday businessman triples himself: referencing his sums, calculating them, and copying down the results, working at a desk somehow both well lit and void of context. The ad's copy suggests that *VisiCalc*'s power lies not merely in simplifying or removing the tedium of calculation but also in redistributing one's labor to a higher realm of meaning: the "what if?" "What if sales dropped 20 percent in March?" "What will happen to our entertainment budget if our heating bill goes up 15 percent this winter?" "What if the oscillation were dampened by another 10 percent?" By redirecting the user's focus toward speculation, *VisiCalc* proffered an opportunity to displace the uncertainty of the unknown into an infinite number of possibly calculable futures—to replace indeterminacy with probability, to develop strategies for every possible outcome. In the shadow of economic unknowns, *VisiCalc* was not a technology for time-saving but a tool that ensured you would always have a contingency plan. Software had never been sold this way: appealing to an individual not on the basis of what made the software good, or useful, or easy but with a sensibility for the way it might architect new hopes over uncertain foundations.

As it turned out, a large part of *VisiCalc*'s popularity among white-collar professionals was that it resolved a problem of access to computing power that Bricklin himself had never set out to solve: the executive's interest in tighter control over figures and greater flexibility in economic projection. It wasn't that financial modeling software, or other kinds of programs for dynamic calculation, hadn't existed before. Software packages specialized for the financial, banking, or retail sector had proliferated within commercial businesses large and small since the 1960s.[79] But within businesses of nearly any size, the access any given employee had to a computer was limited, constrained by the rules and structures of a company's data processing department (essentially the room where the mainframe computer or time-sharing minicomputer hub was kept, guarded by the employees who managed that computer's use).[80] Access was never direct, and the computer was largely an object of mystery. At most companies, in most scenarios, no rationale existed for an *individual* employee to have exclusive access to a computer.

VisiCalc, and the Apple IIs that it brought through the door, redistributed computing power within the workplace. As *BYTE*'s founder, Carl Helmers, wrote in the magazine's August 1979 issue, "The techniques used in Visi-Calc [*sic*] are possible only . . . when the concept of 'one user, one processor' is employed, ie: when the computer power is 'personal.'"[81] Consumer microcomputers were the condition of opportunity for white-collar professionals, especially those in the expanding information industries of the late 1970s, to more closely manage their own research, planning, or budgeting analysis, which was preferable to relying on internal, centralized data processing departments that had little bandwidth for running and rerunning fiscal or scientific scenarios with countless small changes.[82]

So up and down Wall Street, executives went through consumer-facing retail distribution channels to procure microcomputers, creatively accounting their Apple IIs past the data processing departments and into their offices on the company dime. "The check for a computer installation would have the legend 'Furniture' annotated on the stub," wrote the personal computing consultants Barbara McCullen and John McCullen, summarizing the history of these transformations in their early 1980s essay "Screen Envy on Wall Street." "In other words, the firm was purchasing a $9000 desk that just happened to have a funny looking machine sitting on top of it at the time of delivery."[83] In the artifact of *VisiCalc*, a user-centric orientation to software converged with a hardware platform designed for use by only one person.

For professionals whose job it was to massage numbers, manage data, and scrape bottom lines, the direct access to calculation provided by *VisiCalc* was revelatory. "Hours of figuring cost projections eliminated," one computing advocate lauded. "Days of waiting for revised estimates reduced to seconds."[84] *VisiCalc* would be released on other systems in the years following, including later models of the TRS-80 as well as some Commodore machines, but its initial association with the Apple II continued, especially among an expanding entrepreneurial class that viewed Apple Computer as a slickly marketed startup parable, a nimble David facing off against the Goliaths of an outmoded industrial economy.[85] Two years after its release, *VisiCalc* was still outselling its nearest Apple II software competitor by a two-to-one ratio, and it remained in the top ten of all software products sold month to month for the Apple II until December 1983.[86] As one Apple II journalist wrote in December 1981, "The continuing strength of *VisiCalc* in the Apple market is a story so often told that it tends to become ho-hum. It should not be so. *VisiCalc* has validated the personal computer as a useful business tool, and it's the business user who is now flocking into the computer stores everywhere and queuing up for the product."[87]

Business journalism was quick to latch on to the perception of microcomputing as a new trend, with *VisiCalc* as its central star. The program was feverishly covered in mainstream business periodicals, including *Fortune*, the *Wall Street Journal*, and *Inc.* (which began in 1979 as a "magazine for growing companies," the canary in the coal mine on a burgeoning American obsession with entrepreneurship). And a new wave of profitability would be achieved following the 1981 release of the IBM 5150, better known simply as the IBM PC. For business users, the IBM PC was a shot heard 'round the world. If a company's data processing department had been reluctant to invest in microcomputers because of their perceived status as computational toys, IBM's seal of approval solidified the trending perception of the microcomputer as the next eagerly awaited advancement in American business. Thus, despite *VisiCalc*'s ongoing advantageous association with the Apple II, it would eventually sell more copies for the IBM PC than for any other platform—a testament to the fact that, no matter what the headlines claimed, most American businesses were fundamentally conservative in their technological orientation and unwilling to experiment with microcomputers until they were normalized through their association with the largest "traditional" computing corporation on earth.

Once *VisiCalc* had given professionals a taste of what microcomputers could do for them, the scale of the business computing market boomed, developing its own distribution specializations distinct from software development that only targeted consumers at retail. Software publishing for businesses would become the largest segment of the software market by the mid-1980s, dwarfing the consumer and education markets with cumulative unit sales for top products estimated at 2.3 million and revenues near half a billion dollars.[88] The top three business software publishing companies (Personal Software/VisiCorp, Lotus Development, and the recently de-hyphenated Microsoft) exceeded the revenue of the top three consumer software

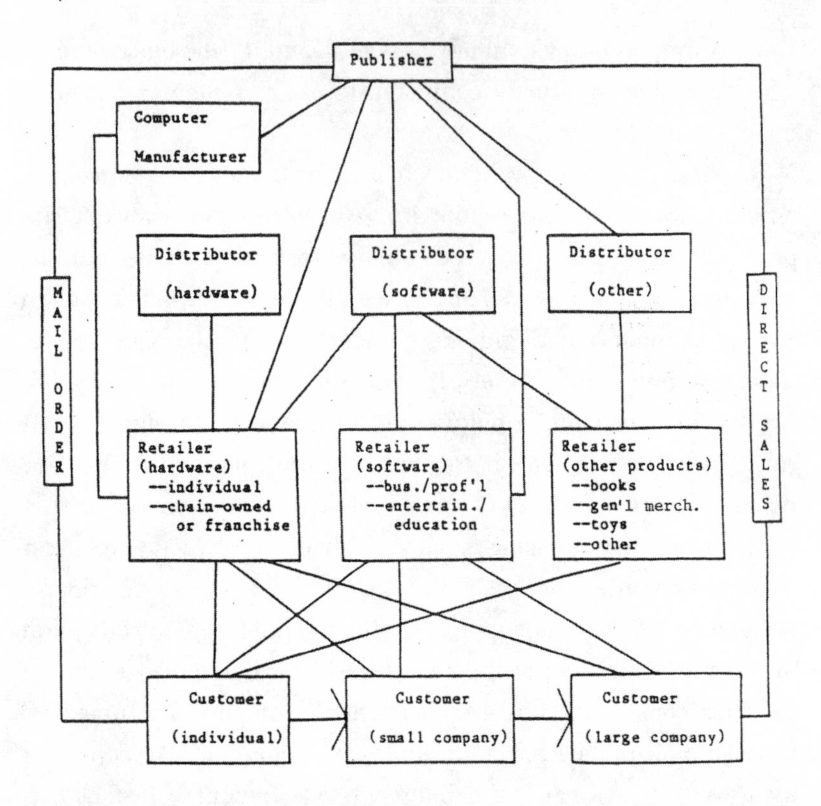

Diagram visualizing "distribution channels for business/professional software." Illustrates the complex pathways through which software made its way to individual consumers and large and small businesses. Published in Efrem Sigel and Louis Giglio, *Guide to Software Publishing: An Industry Emerges* (1984), 31. Image courtesy Efrem Sigel.

companies (Sierra On-Line, Spinnaker, and Brøderbund) by a five-fold margin.[89] Business software would rightfully be considered "*the* major area for software publishers" by the mid-1980s, "the one market where revenues and profits are significant for the leading publishers."[90]

The scale of these margins was due to two factors. First, large corporations' interest in microcomputing was rising, thanks to the IBM PC, and the individualized nature of the microcomputer required these corporations to buy individual units of software in large quantities to outfit entire departments or groups of employees (pushing many business software publishers to hire personnel to handle direct sales and marketing to such corporations). Second, software like

VisiCalc was finally making microcomputers useful enough, and easy enough, to win over small business owners to the idea of owning a computer, giving the business software industry access to a market that comprised an estimated 97 percent of all businesses in the United States. Thus, in just a few years, the market that had been starved for "practical, useful, universal, and reliable" software, as Ben Rosen had lamented in 1979, was transformed, and it was finally helping to bring microcomputers to the masses.

* * *

What *VisiCalc* offered was not a mere computerization of pen and paper practices. Rather, it provided something that had never been possible before: the capacity to *see* the instantaneous transformation of data ripple across rows and columns as the computer refreshed its screen memory. "If you change any of the numerical data, the electronic worksheet instantly displays a new result. Automatically," promised *VisiCalc* marketing. "You can play 'what if' as often as you wish to solve thousands of different problems. When finished, you can get a hard copy of all the information on your worksheet from your computer printer. Absolutely no programming is necessary."[91] Whether it was for a Wall Street stock brokerage or a small-town Main Street storefront, *VisiCalc* promised a way of *assessing* business operations that could be commanded by a single individual.

The sensation *VisiCalc* bestowed—of not just commanding a Cartesian view of a world of numbers down below, but of being able to alter their output—emboldened fiscal manipulators to tweak scenarios to perfection and then execute them in the real world. *VisiCalc*'s fundamental orientation to instantaneous financial calculation would enable new cultures of rapid economic projection, especially within the financial sector, and went hand in hand with an internalization of economic risk, a sort of financialization-of-the-self that *VisiCalc* enabled through the presumed transparency of its economic modeling. "In the past, before spreadsheets, people would've taken a guess,"

Bob Frankston told the tech journalist Steven Levy in 1984. "Now they feel obligated to run the numbers."[92]

What all this amounted to was an epistemology that has sustained into our present moment, a "spreadsheet way of knowledge," as Steven Levy termed it in his November 1984 *Harper's* article under the same name. In this roving treatise, Levy explains the spreadsheet to the uninitiated through a personalized tour of the church of spreadsheet-ology: "There are corporate executives, wholesalers, retailers, and small business owners who talk about their business lives in two time periods: before and after the electronic spreadsheet." Inspired by the fervor of the users he interviews, Levy describes the spreadsheet as the new double-entry bookkeeping, the crisp mastery of oil painting, a transcontinental railroad, what horses were to cowboys—in other words, a new climax in Western models of capture and conquest, a blend of high speed and high detail that conferred on its faithful an "unshakable belief that the way the world works can be embodied in rows and columns of numbers and formulas." Levy tracks this transformation as part of a new "entrepreneurial Renaissance" taking hold in the United States, embodied in "a new breed of risk taker who creates businesses where none previously existed"—and one uniquely reliant on the spreadsheet for its ability to conjure something out of nothing.[93]

Levy's diligent analysis also reveals a historical arc in the uptake of spreadsheets. While their initial appeal might have been in saving time, their accelerated use also produced new forms of work. Quarterly updates could be replaced with monthly, weekly, or daily updates, allowing for continual instantaneous assessment. But the real delirium of *VisiCalc*'s appeal ultimately returned to the "what if" factor—the fact that spreadsheets, as one executive put it, allowed them to experiment with "a phantom business within the computer":[94]

> All this powerful scenario-testing machinery right there on the desktop induces some people to experiment with elaborate models. They talk of "playing" with the numbers, "massaging" the model. Com-

> puter "hackers" lose themselves in the intricacies of programming; spreadsheet hackers lose themselves in the world of what if. . . . The experiments . . . are far-flung attempts to formulate the ultimate model, the spreadsheet that behaves just like an actual business.[95]

The presumably god's-eye view afforded by spreadsheets, users' ability to pinpoint and cast off low-performing assets (or low-performing employees), was fuel for financial operatives hell-bent on finding new pathways for capital accumulation among economic elites in an uneven economy. This was "personal" computing in a new way: computer power harnessed to serve an individual's personal anxieties and ambitions.

Perhaps one of the most iconic exemplars of this tendency was the investment banker Michael Milken, the Wall Street "junk bond king," who made a fortune instigating high-yield leveraged buyouts of companies experiencing depressed stocks during the economic ups and downs of the 1980s. *VisiCalc* was essential to Milken's process, allowing him to simultaneously stalk multiple companies, calculating which companies would make good targets for buyouts based on their cash flow relative to the amount of debt they could bear. As the historian William Deringer has written, "Spreadsheet software like *VisiCalc* and *Lotus 1-2-3* became a durable augmentation for financial agents" like Milken, who exemplified newly emergent neoliberal economic perspectives about the value of "information" as a commodity unto itself.[96] Milken didn't make money by harnessing economies of scale or pursuing new efficiencies but by manipulating relationships of revenue, debt, and tax write-offs to consolidate interests and restructure companies in favor of the investors he represented. Enabled by the electronic spreadsheet, Milken produced profitability for shareholders (rather than employees) where it previously had not existed.

VisiCalc did not somehow single-handedly create the "Deal Decade" of the 1980s, as it was named by an eponymous book, nor was it responsible for the "shift in emphasis from production to finance as the centerpiece of capitalist class power" that marked American economic trends from the late 1960s on.[97] However, it is necessary to understand that *VisiCalc* existed in a symbiotic relationship with these larger trends. As Kera Allen has explored in her examination of *VisiCalc*'s use by the Rockefeller Foundation to address agricultural shortfalls in Tunisia, the program encouraged modes of data analysis that were largely only about their own propagation, a scenario in which stakeholders mistook the "amassing data for a more complete analysis" for "improved analytical capacity."[98] Like the "what ifs" *VisiCalc* proposed for individual users, the very existence of the program invited a different kind of "what if": What if the struggles of the American economy were not a challenge to America per se but an opportunity for continued innovation, a time for necessary upheaval and the shedding of old skin? Such at least was the party line of the cult of entrepreneurship, which found ready circulation for new ideas among an American mass media eager for a hopeful spin on uncertain times. In this way, the economic upheavals and worries of this historical moment helped shape what "personal" computing would mean for decades to come: the start of a long trajectory of computer use built on the belief that—given enough data, enough processing power, just the right model, just one more innovation—computers could help us massage our future into whatever we wanted it to be.

* * *

For all the advantages of visibility that *VisiCalc* afforded, there would always be corners of the world it couldn't calculate. For example, nothing in its rows and columns could have predicted a $60 million lawsuit brought by *VisiCalc*'s publisher, VisiCorp (Personal Software had renamed itself in 1982 as part of a "high visibility and aggressive marketing program to establish brand recognition"), against *VisiCalc*'s developer, Software Arts, in September 1983.[99] The lawsuit was the culmination of long-brewed tensions, as organizational growth at both companies pulled apart the once-aligned interests of Fylstra and Jennings and Bricklin and Frankston. Communication crumbled; accusations grew bold. VisiCorp was overshadowing Software Arts in taking credit for *VisiCalc*. Software Arts wasn't responding quickly enough to improve the product, causing VisiCorp to lose market share on the IBM PC. Neither party was satisfied with the royalty deal struck four years earlier. Bricklin and Frankston's company had been paid over $22 million in royalties by early 1984. The sheer amount of money VisiCorp was ceding to Software Arts became an inescapable source of friction between the two companies—especially since, no matter how many complementary products VisiCorp released, *VisiCalc* brought in the majority of VisiCorp's revenue.[100] Without a diversified revenue stream, VisiCorp's profitability rested on a single piece of software. The testy machinations between VisiCorp and Software Arts, full of suits and countersuits, became a case study on "how a software winner went sour," as one *New York Times* headline put it.[101]

In the shadow of all this upheaval, something else *VisiCalc* couldn't have predicted was changing the stakes of business software: Mitch Kapor, a former VisiCorp employee, had developed a not so little program called *Lotus 1-2-3*.[102] Released in spring 1983 and customized especially to the new 16-bit IBM PC, *Lotus 1-2-3* harnessed features *VisiCalc* had yet to implement. With better features and product integration, *Lotus 1-2-3* set itself apart from *VisiCalc* in the eyes of corporate information technology services executives eager to make the right choice for their company. VisiCorp and Software Arts were bickering away in court as Kapor ate the business software market for lunch. In the end, it was all cannibalized: Software Arts' remaining assets were sold to Kapor's company in spring 1985—a final bid to stave off bankruptcy. With no reason to keep a prime competitor alive, Lotus shuttered *VisiCalc*. As for VisiCorp, despite being the world's fifth-largest microcomputer software company, bringing in $43 million in sales in 1983, the fall of *VisiCalc* sent the company into a rapid decline.[103] By November 1984, VisiCorp was up for auction.[104]

But even with *VisiCalc* gone, the "spreadsheet way of knowledge" was here to stay. What *VisiCalc* showcased, in a way no program had before, was how computational power might be leveraged for individual gain. The value-add was in black and white (or green and black or orange and black, depending on your monitor). The program changed how people thought about computation—not just about its elasticity and ease and accessibility, but about its presumed transparency, its hypothetical objectivity, the way it transformed the problem of economies big and small into just another numbers game.

While the Apple II, as we have seen, was explicitly designed to appeal to average consumers, not just programmers and hobbyists, it took a program like *VisiCalc* to turn those potential consumers into *users*: to scope computational power to the level of individual anxieties and desires. The brokers and hedge fund managers and middle-tier executives who leveraged *VisiCalc* most aggressively weren't everyday people by any stretch, but they were influential and persistent in their pursuit of financial gain. In *VisiCalc*'s cascade of immediately updated rows and columns, they saw what they could *do* with a computer of their own. And the effects were trickle down. *VisiCalc*'s proposition was that efficiencies could be found anywhere, even in the smallest small businesses. *VisiCalc* subtly reframed the value of the personal computer from a proposition of ownership to one of agency: with the right software, in the right user's hands, the microcomputer was not just property, an object to tinker with, but a way to gain power over a complex world. This proposition was powerful enough, compelling enough, to turn the tide on the Apple II itself—to help transform it, from just another competitor in the field, to *the* computer of the early 1980s.

4

Games

Mystery House

Why would anyone pay for a computer game? Not a console game or an arcade game, mind you, but a *computer* game? Before the commercialization of the microcomputer, computer games had always existed in a world apart from economic incentives. They were developed for fun by programmers, hosted freely in the subdirectories of mainframes and time-sharing systems, or passed across the government-funded, proto-internet telecommunications systems that connected university research hubs in a growing network. The free movement of such software from site to site, the ease with which a programmer might covert a preexisting game to a new system or modify code to expand playability—all with little care for parochial concerns of authorship or ownership—was a hallmark of game development in the era of large and midrange computing systems that dominated the 1960s and 1970s.

So computer games, like almost everything else that happened in such institutional computing sites, were part of a closed world. Everyday people rarely had access to the games created by computing insiders. Furthermore, general-purpose computers were too rare, too expensive, too important, to be used for dedicated entertainment purposes.[1] And even if some enterprising nonprogrammer *had* stumbled upon a time-sharing system, could make sense of subdirectories, or figure out how to execute a program? They probably wouldn't have been impressed with what they played, not compared to the colorful graphics, blippy noises, and real-time interactivity that typified the emerging arcade and console game industries beginning in the early 1970s.

With a few notable exceptions like MIT's outer-space dogfight *Spacewar!*, computer games were mostly text-based, black-and-white affairs played by typing commands into teletype terminals. The games that existed in this form were most often adaptations of existing sports, board games, or card games—everything from mancala to baseball to blackjack existed in computerized form—or strategy simulations that played to the strengths of number-crunching machines, games like *Hamurabi*, *Civil War*, *Star Trek*, even the now-iconic *Oregon Trail*. Often understood as puzzles, simulations, ludic experiments, or small, novel entertainments rather than "games," these programs were largely afterthoughts, not unlike a crossword puzzle kept on the tank of an office commode: you played when you had a little time on your hands, to take back the time your boss took from you, but you didn't put much thought into the whole affair.

Computer games functioned like this because there was no market for selling games when such a small fraction of the population had direct access to computing power only at their jobs. And what boss wanted their computer users playing games at work? Microcomputers like the Apple II, however, changed these dynamics. If an Apple II meant that computing power was now a private commodity rather than a shared resource, then software could, and did, follow suit. While many of the first games released for microcomputers were simply conversions and adaptations of programs that had long existed on mainframes and minicomputers, the opportunity to *sell* game software produced an economic incentive for the development of more complex and sophisticated games.

We might begin the story of why game software mattered from any number of launch points; unlike *VisiCalc*, there was no singular program that set the bar for everyone else. But let's begin somewhere games rarely dwell: the home. Or in our case, "THE FRONT YARD OF A LARGE ABANDONED VICTORIAN HOUSE. STONE STEPS LEAD UP TO A WIDE PORCH." Such is the opening description of *Hi-Res Adventure: Mystery House*, a game launched in May 1980 by the husband-and-wife team Ken and Roberta Williams (see pl. 11). *Mystery House* is most frequently remembered today for its firstness: as the first adventure game with graphics, as the first game released by what became one of the most iconic computer game companies in American history, Sierra On-Line, and as the first computer game ever designed by Roberta Williams (also one of the first female game designers in the world).[2]

Yet what made *Mystery House* a marvel in its moment wasn't the whodunit plot line, the construction of its puzzles, or the historical accolades that could not yet be known. Rather, it was the game's technical accomplishments: *Mystery House* had more than seventy black-and-white line illustrations, a seemingly impossible feat given that the entire program was stored on a single 5.25-inch floppy disk. *Mystery House* would be one of the first commercial games, alongside *FS1 Flight Simulator* and Bill Budge's early works, to move beyond the tradition of simply adapting popular mainframe and minicomputer software and explicitly exploit the hardware limitations of the Apple II as a platform. The attraction of the game's novelty is attested by its almost immediate popularity, raking in tens of thousands of dollars in the months immediately following its release and holding its position in the top thirty of all Apple II software sold for up to a year.

Mystery House serves as our entry point into long-forgotten dimensions of early commercial computer game development. In the Williamses' cottage operation, we find another model of how microcomputer software businesses emerged in the moment when no one knew what it meant to be in a microcomputer software industry. Rather than split responsibilities between a developer and a publisher, like Software Arts and Personal Software, the Williamses began as developers who grew into publishers—something they were capable of doing only because of the low barrier to entry that existed for entertainment software. The Williamses' collaborative dynamic is also worthy of examination, insofar as they embody perhaps the earliest known separation between the roles of programmer and designer in game development (a division of labor rare during the early to mid-1980s but now endemic throughout the industry). And beyond these biographical details, *Mystery House*'s development process highlights the technical qualities that made the Apple II a particularly good platform for game production, especially with regard to the system's unique "hi-res" graphics mode, as well as the significance of Apple's release of the updated Apple II Plus in summer 1979, which helped stabilize floppy disk peripherals and 48K memory as the platform's baseline standard.

Whatever its notability, however, *Mystery House* was just one of hundreds of games released for the Apple II. In the fast-moving microcomputer software industry, games moved the fastest, largely owing to their status as discretionary entertainment. No sooner was the game released than Ken Williams began working on a second adventure game loaded with even more impressive graphical tricks—a cycle that would soon come to dominate the nature of game production overall. Thus the computer game category is especially illustrative of the dramatic churn and underlying precarity of the microcomputer software industry more broadly. Following *Mystery House*'s lead, I will use the game's boom and bust, alongside the Williamses' need to keep providing new marketable content, as a handrail for understanding the convoluted and sometimes contradictory trajectories game development underwent during the early to mid-1980s. As I show, the bursts and fizzles of the popularity cycle, the proliferation of game genres, and pressure from console and arcade manufacturers all rapidly shifted what little institutional knowledge existed in the early American computer game industry.

While other microcomputers, like the Atari 800 (1979) and the Commodore 64 (1982), also attracted a vibrant and prolific community of game developers, the Apple II's longevity makes it a superior object for understanding the growth of the computer game industry as a whole. Preceding both of these competitors, the Apple II gives us a clearer sense of what industry practices were like at their earliest moment. Furthermore, Apple II game software developers had to compete in a denser market of home and business software products—thus allowing us to better understand how games performed not only against one another, but within a larger consumer software ecology.

The history of the game industry has largely been told from a consumer perspective, insofar as entertainment software across arcades, consoles, and microcomputers is often presented as an undifferentiated chronicle of an internally coherent industry. Yet from the point of view of entrepreneurs like the Williamses, microcomputer game software was a vexing category. At the industry level, computer game software had to function like any other bit of code sold on a floppy disk, moving through the same distribution networks, occupying the same retail sectors, and being legible as a consumer application. Yet such software also had to function recognizably *as a game* during the first great boom of American video game consumption, existing alongside the faster, brighter, and louder experiences circulated on arcades and consoles. *Mystery House* thus emerges at the intersection of computing and gaming, expressing tensions inherent to all computer game production at this time.

Within a short span of years, microcomputer games transformed from derivative novelties to original creative works expressly designed to show off, in a way no other kind of software could, what a computer could *do*. Using software to bend and exploit the limitations of hardware, in uniquely entertaining and interactive ways, is an underrated component of what made games important as a category of software at the dawn of personal computing. Games may indeed have been marked by their novelty and ephemerality, but as we shall see—and as *Mystery House* helps us explore—that was often exactly the point.

* * *

Roberta Williams was not going to be dissuaded. She "blindsided" her twenty-six-year-old husband, Ken, at their favorite local steakhouse, or at least that's how Ken remembers it—Roberta leaning in with an air of insistence he wasn't accustomed to. It was the early months of 1980, and she wanted him to help her make a computer game. Her pitch held all the tropes of an Agatha Christie–style domestic horror, where every turn around a hallway or opening of a door might reveal another body. She described the scenes to Ken with escalating enthusiasm, loudly enough that other diners took notice. This wasn't the first time Roberta had made this appeal, but it was likely the first time Ken *listened*, closely and carefully, to his wife's proposition. This wasn't something just to do for fun. This was something to do for profit.[3]

Neither Ken nor Roberta fit the profile of the computer entrepreneurs we've seen thus far. They had no Ivy League pedigree, no fathers employed on cozy government engineering contracts, no advantageous social milieu. Both born in the early to mid-1950s, Ken and Roberta grew up mostly in the suburban outskirts north of Los Angeles. Raised on the lower fringes of the middle class, Ken was a consummate hustler from an early age, earning what he could from odd jobs while teaching himself how to fix basic consumer electronics (thanks to a father who worked in television repair).[4] As for Roberta, she recalls her teenage self as dreamy and undermotivated and fixated on boys. Neither involved themselves much in the grander scale of political events unfolding around them during the 1960s and 1970s. They met through mutual high school friends, settled into a relationship, and married before either was twenty years old.[5]

While Ken may have tinkered with electronics as a boy, he wasn't privileged with childhood access to computing. Ken's first interaction with a computer didn't come until college, at Cal Poly Pomona, where he took a large-systems FORTRAN programming class. He had been enrolled as a physics major, but when Roberta became pregnant, Ken decided to expedite his career path: he dropped out of college and signed up for a nine-month trade school program in electronic data processing. For the next half decade or so, Ken worked in the world of large-systems programming, bouncing around gigs, salary hunting, forever picking up new tricks, new languages, even taking on freelance work in the evenings, coding late into the night on a teletype terminal he kept in the house.[6] Roberta also contributed to the family war chest, flipping their houses while caring for the children, and even did brief stints in the computing industry—working as a computer operator changing out magnetic drum storage and doing light COBOL programming, though she had no passion for the work.[7] The two were hungry, ambitious, if in different ways, but there should be no mistaking the fact that what Roberta pitched Ken across the table of a steakhouse was a product *and* a game.

Despite not being particularly enthusiastic about computing generally, Roberta had gotten hooked on computer games late in 1979, starting with her exposure to the text-based narrative game *ADVENT* (better known in game history's canon as *Colossal Cave Adventure*), which she first played on Ken's teletype terminal.[8] Computer games weren't a new concept to Roberta. She had watched Ken play traditional mainframe games on the teletype before, little simulations like *Star Trek* and other such programs, but *ADVENT* was something different. Unlike the repetitive statistical simulations common to large computing systems at the time, *ADVENT* was a text-based treasure hunt spread out across a modest interactive world. Players navigated space one "room" at a time, like traversing a grid in which each square had its own description, objects, entrances, exits, puzzles, and enemies.[9]

If not quite a story and not obviously a game in the usual sense, *ADVENT* certainly had tension and drama, challenge and humor. Perhaps more important, it also had affordances that were legible to Roberta. Not only was the game's interface relatable, given its use of natural-language, verb-noun input syntax, but something about the game's narrative contours and engaging puzzle design gripped her. "I just couldn't stop," she said. "It was compulsive. I started playing it and kept playing. I had a baby at the time, Chris was eight months old; I totally ignored him. I didn't want to be bothered. I didn't want to stop and make dinner."[10] Roberta was experiencing, in stark terms, the kind of cognitive stickiness, the "cathexis," so popularly documented by players of early digital games.[11]

It would take a microcomputer, however, for this game-play experience to be anything more than an isolated event happening across a time-sharing network accessed in the spare bedroom of the Williamses' home. In keeping with their entrepreneurial instincts, the Williamses were entertaining the prospect of buying a microcomputer that Christmas, chiefly to support Ken's ambition of developing a marketable FORTRAN compiler.[12] They had some exposure to a TRS-80 but ultimately settled on an Apple II, which they brought home early in 1980.[13] Thus it was through the unique affordances of at-home computing that Roberta discovered the world of software-as-commodity.[14] The microcomputer became her pathway to more games like *ADVENT*, more story-world treasure hunts set in fantastical realms and demanding her clever engagement. But most of the games she found were just *ADVENT* clones; little original work existed. If the teletype terminal in their home gave the Williamses access to an unpolished world of freely circulated software, the Apple II was a vortex for considering the relationship between individual software production and capital accumulation.

In a matter of months, then, Roberta Williams went from never having played a computer game to having played nearly every entrant in the adventure game genre. This allowed her to imagine what else might be possible—and more important, sellable—on the Apple II. Her perspective would also have been shaped by Ken's latest side hustle: he had begun working as a software distributor soon after getting the Apple II, going door-to-door selling products to small regional computer shops out of the trunk of his car.[15] As a couple, they would have known what was out there, what was selling, what moved units. With a few exceptions, game software for the Apple II was underwhelming and derivative, despite the platform's potential. Not enough Apple IIs had been sold, not enough programmers were developing for the machine, for much experimentation to have coalesced into products by early 1980. Roberta's aspiration to design a game was thus not a flight of fancy but grounded in a surprising amount of practical knowledge, not to mention tactical acumen, for a woman who otherwise had little interest in the bits and bytes of hardware and software.

Roberta had no interest in programming herself, but her experience working in COBOL, as well as the ambient presence of computing in her domestic life, gave her a programmatic literacy essential to scoping a project that would be feasible on an Apple II. Her primary inspiration for setting and story have been attributed to the board game *Clue* and Agatha Christie's *And Then There Were None*, while for the natural-language input, object gathering, and puzzle proceduralism she had gleaned lessons from the other adventure games she played.[16] Since she was not interested in developing her work on the computer, her prototypes were physical and were created, as the story goes, at her kitchen table—a choice of setting undoubtedly influenced by the fact that she was also taking care of two young children. The domestic context of *Mystery House*'s production poetically refracts the game's content. Whereas most games in that historical moment were occupied with outer space, fields of physical play, or fantasy castles and dungeons, *Mystery House* shrugged aside the inspirations that dominated male computing culture.

By the time Roberta "blindsided" Ken at the steakhouse, she knew what she wanted and what she thought could make the game better than anything else that was currently on the market: images. Roberta sought both immersion and novelty and valued illustration over description. Importantly, she had a sense of Ken's talent with programming. Roberta Williams may have been a born storyteller who felt compelled to create, as some biographical accounts spin the tale, but she also had confidence that her design had commercial potential.

* * *

In 1980, the only consumer-grade microcomputer capable of running a program with detailed images, like those Roberta Williams intended for *Mystery House*, was an Apple II with Disk II peripherals and 48K RAM. Thus *Mystery House* is uniquely tied to the platform's expansive graphical features, as well as the presence of floppy disk peripherals and a broadening base of Apple users outfitted with 48K memory. In this sense, *Mystery House* rode the cutting edge, showing off its system hardware to the fullest known advantages of the time.

While games were a popular form of software production on all microcomputer systems, the Apple II's capabilities made it especially apt for those programmers looking to press the limits of what was possible on a home computer system. After all, Steve Wozniak had designed it that way.[17] An avid arcade gamer himself, Wozniak tricked out the Apple II with features that were widely appealing to those who wanted to play—or make—games and had been inspired to do so by his own experiences programming games in hardware.[18] These features included support for color graphics and a high-resolution graphics mode, as well as minimal audio capacities and the two game paddles, or "one-dimensional joysticks," packaged standard with every Apple II.[19] Wozniak's choices were due more to his personal sensibility for what made computing fun than to deep consideration of what would make the system versatile for third-party developers, yet the result was advantageous for Apple all the same: more programmers attracted to making software for the platform.

The Apple II's graphical capabilities especially set the system above its competition in 1977—even if the immediate applicability of these features was initially underutilized in commercial software production. The most straightforward of these features was the Apple II's capacity for 16-color graphics when used with a color monitor or color television, which was a direct extension of Wozniak's desire to approximate arcade-like game experiences on the platform.[20] Neither the TRS-80 nor the Commodore PET ever provided color graphics support.[21] But the Apple II's commitment to versatility in graphics production went beyond merely color display. It also offered two resolution modes: a low-resolution, 40 × 40 pixel graphics mode supporting the promised sixteen colors, and a high-resolution, 280 × 160 pixel graphics mode supporting four colors.[22] "Resolution" is a measurement of how many pixels, or the smallest controllable element of a picture on a screen, a computer can address. In microcomputers, resolution was typically constrained by the computer hardware, not by the monitor—which is exactly what enabled Wozniak to design the Apple II with two resolutions on the same monitor. The distinction between resolution and color selection was one of the trade-offs Wozniak configured to make high-resolution mode possible: low-resolution mode, which produced chunkier graphics, could use a wider array of colors, while high-resolution mode "afford[ed] a larger canvas allowing for more detailed renderings, at the expense of having fewer colors available."[23]

This extraordinarily high resolution, despite reduced color options, allowed Apple II computer games to offer detail greater than anything else available on the market for several years. But "hi-res," as it was often written by journalists, was also a memory hog: a hi-res screen required 8K of memory to express, in addition to whatever memory was needed to run program instructions. Thus, hi-res was uncommon in very early Apple II games, given how expensive even small units of memory were in 1977. The commercial imperative to have one's games appeal to the broadest possible market of users led to game designs that privileged small memory requirements and helps explain the general market tendency toward games with low visual interactivity between 1977 and 1979.

It would not be until 48K memory became standard, buoyed by the release of the Apple II Plus in 1979 and further accelerated by the phenomenally popular 48K *VisiCalc* the same year, that more programmers began experimenting with the hi-res mode in commercial game production. The Apple II Plus was developed chiefly to resolve inadequacies in the original Apple II's BASIC program, but Apple Computer also took advantage of lower memory costs to raise the memory threshold of the system to 64K by removing accommodations for smaller memory options.[24] Consistent with the exponential reduction in memory prices throughout the 1960s and 1970s, the results for the Apple II were dramatic: the Apple II Plus came standard with 16K, yet cost $100 less than the original 4K Apple II from just two years earlier.

Given these cost efficiencies, many consumers elected to buy the 48K Apple II (the minimum required to run *VisiCalc*), getting a truly high-end machine (for the time) at the cost of only $1,500 in 1979.[25]

As 48K became the baseline memory for an expanding community of Apple II users at the beginning of the 1980s, this emergent standardization became the condition of possibility for a vibrant world of game production. Low-res games dropped from the market almost entirely; only the simplest productions, or games lacking graphics altogether, failed to use the Apple II's hi-res mode. This spawned a hearty competition among programmers, as they sought to outdo themselves and one another, to bend hi-res graphics to their will. Over the next several years, programmers successfully got the Apple II to produce hi-res effects the system was theoretically incapable of—supporting more than four colors, inserting text at other locations on the screen, and many other flashy, graphics-oriented techniques.

This was the technological, economic, and creative software ecosystem within which *Mystery House* emerged. The Williamses were not so ambitious as to attempt color graphics, but the Apple II's hi-res mode was essential to depicting the linear perspective and foreshortened interior spaces that dominate *Mystery House*'s setting. It also allowed for the depiction of specific objects that could not have been recognizably rendered at a lower resolution: a piece of paper on the floor marked "NOTE," a candlestick left on a dining room table, a flower ominously abandoned at a murder scene. Whereas earlier text adventures demanded close attention to written clues, a kind of reading between the ludic lines, *Mystery House*'s illustrations were puzzles unto themselves, decodable only through their own modest visual logics.

In prototyping *Mystery House* with a wide array of images, Roberta's designs presented Ken with a technical challenge. A 5.25-inch floppy disk could not hold many images because image files required color and position data to be stored for every pixel, resulting in large file sizes (even today, this is all a digital photo is: a vast table of specifically organized color data). Theoretically, the Williamses could have chosen to store many images across many floppy disks, but that

Game screens from *Mystery House* (1980), including the game's starting screen. Screen captures by the author.

would have made the game prohibitively expensive to manufacture and onerous to play, insofar as swapping out floppy disks was a time-consuming activity. To make the game viable on the market, and not just a novelty imagined by Roberta, Ken was stuck with figuring out how to fit the entirety of *Mystery House*, including all of Roberta's desired images, on a single 5.25-inch disk. Lacking utilities for image compression, Ken needed to come up with a different way of accommodating the images essential to *Mystery House*'s value proposition. Thus it was the distribution format, not the Apple II, that forced the technical hack that made *Mystery House* a viable product.

Ken Williams's solution was not to program the game's illustrations as image data at all but to render them as *instructions* for drawing lines between on-screen coordinates.[26] Instead of having to store image data for every pixel, all Ken had to do was store the X and Y coordinates for every point, along with the subroutine that drew the lines between these points. To capture coordinates for the drawings, the Williamses bought a VersaWriter, a vector-based drawing tablet with a drafting arm and stylus meant for tracing over images. The images were physically drawn by Roberta herself, then traced using the tablet.[27]

But even this approach had its limits. Ken recalls, distinctly, haggling Roberta's expansive visual demands down into the range of technical feasibility. According to Ken, Roberta initially aspired toward "grandiose visions of having 100 locations," which Ken negotiated to the game's more manageable seventy-some screens. The process was one of constant push and pull between the couple. As Ken explained:

> My guess is it was back and forth, and I had the idea for how to digitize pictures and she probably had a picture that was way more complex than I could fit in. I told her, "Somehow you got to get a picture down to no more than 75 points because then 4 bytes per point, that's 300 bytes." I wish I had better memory of this, [I'm] 99 percent certain I would have given her . . . that's how she functions best, is when you just say, "OK, here's your parameters, you have 50 points per picture and that will average out to 300 around one picture and try to pick another one and go simpler and go under it."[28]

The co-constitutive relationship between Roberta's demand for what she felt would provide greater sensorial immersion and Ken's negotiation of technical constraints produced a paradoxical state of affairs: Roberta, the member of the team with lesser technical experience, was the one largely responsible for instigating the technical innovation of *Mystery House*.

The creative and technical dynamic between the couple also produced a distinct organization of labor. Roberta's disinterest in programming and Ken's disinterest in the details of the game's content induced a separation between the programmer and the designer that was unique to game development at that time. Typically, microcomputer games had solo developers—the young, bleary-eyed hackers so often represented in film and television who created games as a way to test the limits of their hardware (and themselves). For many developers involved in game software production in the late 1970s and early 1980s, the point of crafting games was to master code, and a way to master code was to make games. The notion of game development as a set of creative and conceptual skills separate from the construction of the game itself did not yet exist. Yet Roberta and Ken Williams subverted this expectation, foreshadowing future divisions of labor that would accelerate within the game industry beginning in the mid-1980s.

* * *

Mystery House wasn't just any game. It was a *computer game*, and in that sense the economic and industry factors that shaped its manufacturing, marketing, distribution, and consumption had more in common with a software product like *VisiCalc* than an Atari cartridge game like *Pitfall*. While popular histories of games have largely lumped together the arcade, console, and computer game industries into an undifferentiated narrative about creativity, fun, and technical innovation, each gaming platform had its own development modes, manufacturing requirements, distribution pathways, economic models, and technical limitations.

Our experience of game play today is increasingly platform-agnostic, allowing us to move seamlessly from smartphone to television to computer, but the game industry of the late 1970s and early 1980s existed in a different technological reality. Arcade games were part of the larger coin-operated amusements industry, making their money quarter by quarter at storefronts and arcades that leased these machines from companies specializing in coin-op maintenance and distribution; individuals, and even stores and arcades, rarely owned the arcade cabinets they hosted. The dedicated hardware of arcade games, each customized to the specific game it was designed for, permitted the dazzling colors, sounds, and movements so commonly associated with the arcade era—but also meant the machines could only do the thing they were programmed to do.[29] Console games, in contrast, were sold as toys or household entertainments. Their appeal was in bringing the magic of the arcade into domestic space (and in doing so, neutralizing parental anxieties about the arcade as an immoral, gendered, and racialized space).[30] Home consoles like the Atari VCS, the Fairchild Channel F, the Mattel Intellivision, and the Bally Astrocade relied on razor-and-blades economics, in which the software cartridges, not the consoles themselves, were the primary revenue stream.[31]

But computer games were different, especially games for a higher-end microcomputer like the Apple II. Few consumers bought microcomputers simply to play games. In 1980, even the lowest-grade TRS-80 was two to three times the price of a video game console.[32] Computer games were a value-add for microcomputer owners but rarely a primary motivation for purchase. Thus microcomputer game software developers understood that their success was not tied to the larger video game industry, which was booming in the late 1970s, but to the overall acceptance of the microcomputer as a domestic technology, which was trending well but had by no means become a dominant fixture in American households.

These economic and industrial considerations inevitably had an impact on how the Williamses sold their software. The very fact that they were able to design a game for the Apple II, and imagine selling it themselves, was one benefit of doing software development for microcomputers. In the arcade and console industries, software development was heavily overseen by hardware manufacturers. Arcade games were designed *as* hardware, meaning development relied on access to expensive electronics components and required substantial financial backing. With the exception of the American startup Atari, most arcade manufacturers in the United States and Japan were experienced companies with long histories in coin-op entertainment or consumer electronics. By and large, even hackers and hobbyists didn't make their own arcade games. Likewise, the console industry initially treated platform design as a trade secret, ensuring that only the hardware manufacturer itself could develop software for its system (an essential component of the razor-and-blades model at the time).[33]

The development ecosystem for microcomputer software was more open, especially for the Apple II, because a microcomputer and its software presented a fundamentally different economic proposition from an arcade machine or a console. Unlike the dedicated hardware of consoles and arcade units, where the computational processing power was designed for games only, microcomputers embraced a tremendous variety of use cases. While this made them more complex to use and harder to sell—computers weren't "plug and play" for decades, no matter what the advertisements claimed—it was also the source of their commercial appeal.

In order to do anything with the computer, however, systems needed software, and it would have been impossible for any microcomputer hardware manufacturer to predict all possible software needs. While first-generation microcomputers, like the Altair and the IMSAI, embraced open architecture and third-party development as a vestige of their hobbyist origins, second-generation systems like the Apple II carried that impulse forward as a market proposition, recognizing that third-party software development enhanced the value of the system. Apple's hands-off approach with its developers—indeed, its willingness to embrace and support companies with innovative products like *VisiCalc*—was what allowed the system's software development world to flourish, further stimulating hardware sales.[34] Developing for the Apple II made the Williamses free agents: they owed Apple nothing, not even a licensing fee, and were able to make whatever economic decisions best served their own interests.

The Williamses did not initially intend to sell *Mystery House* themselves. They understood the overhead involved, the sheer breadth of labor that would be very different from what it had taken to develop the game. The matrix of factors guiding their decisions—an evaluation of time and startup cost risks versus long-term gain—is emblematic of the calculus undertaken by many cottage-industry developers who chose to work with a publisher for modest royalties rather than do the publishing themselves. The Williamses shipped *Mystery House* to Apple first, hoping that the program's standout use of hi-res mode would entice the company to license the product for sale as part of Apple's own software offerings (by the time Apple got back to them, the Williamses had already gone into business for themselves). They also approached Programma International, one of the few major American microcomputer software publishers aside from Personal Software. Programma offered a 25 percent royalty for the game, but the Williamses were unimpressed by this offer, so they made the bold but not unreasonable decision to publish the game themselves. Given the low barrier to entry for marketing one's own game, even modest success was likely to net the Williamses more money than the royalty arrangement, and they were confident of *Mystery House*'s potential appeal. As had always been the case with the couple's can-do hustling style, they were more than ready to trade hard work for marginally increased financial gain.

The Williamses' first step was to develop a marketing plan for the game. The most traditional outlet for this was computer hobbyist magazines, given that the installed base of any microcomputer system was still too small to warrant advertising on television, radio, or mainstream newspapers and magazines.[35] Readers of computer hobbyist magazines were also the most likely to already own and be interested in microcomputer software. The price of such advertising in niche periodicals was also relatively cheap—a fact that the Williamses would use to their advantage.

Everything about the Williamses' first advertisement was tuned to give the impression that their business was far more substantial than it really was—all part of the confidence game of getting strangers to mail them a check for $24.95. At the time, only the largest software publishers took out full-page advertisements to push their wares; independent developers typically purchased discrete quarter-page ads, or even smaller. Yet the Williamses bought a full page of advertising

NEW APPLE II / APPLE II PLUS SOFTWARE FROM ON– LINE SYSTEMS

TIRED OF BUYING GAMES THAT BECOME BORING AFTER A FEW HOURS OF PLAY? ON—LINE SYTEMS IS DEDICATED TO DELIVERING SERIOUS SOFTWARE FOR THE DISCRIMINATING GAMESMAN. THESE PRODUCTS HAVE BEEN SIX MONTHS IN DEVELOPMENT AND PROVIDE THE QUALITY AND SPEED POSSIBLE ONLY THROUGH MACHINE LANGUAGE!

ALL NEW

HI-RES ADVENTURE ("MYSTERY HOUSE")

What is an adventure game? According to the dictionary, an adventure is a hazardous or daring enterprise; an exciting experience; to risk, hazard, to venture on. One who goes on an adventure is a venturer. A seeker of fortune in daring enterprises; a speculator. In essence, an adventure game is a fantasy world where you are transported, via your own computer. You are the key character of the fantasy as you travel through a land the likes of which you will find in books that take you, through your imagination, to the world it is creating.

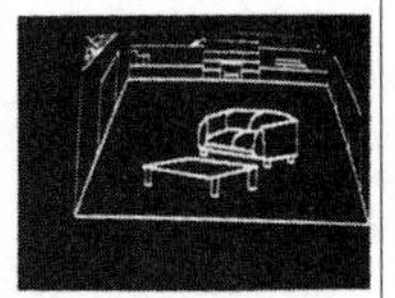

Through the use of over a hundred Hi-Res pictures you play and see your adventure. You communicate with HI-RES ADVENTURE in plain english (it understands over 300 words!) All rooms of this spooky old house appear in full Hi-Res Graphics complete with objects you can get, carry, throw, drop, or ?.

In this particular HI-RES ADVENTURE game, you are transported to the front yard of a large, old victorian house. When you enter the house you are pulled into the mystery, murder, and intrigue and can not leave until you solve the puzzles. Your friends are being murdered one by one. You must find out why, and who the killer is. Be careful, because the killer may find you! As you explore the house there are puzzles to be solved and hazards to overcome. The secret passage-way may lead you to the answer.

ALSO NEW FROM ON—LINE SYSTEMS

SKEET/TRAP have become Olympic shooting sports and and obsession among Scatter-gunners all over the world. These games are the All-American although they have become international.

SKEETSHOOT allows one to five shotgunners to test their marksmanship as they fire from the eight prescribed positions on an official NSSA skeet firing range. Each position provides a new perspective of the field with the pigeons travelling at different angles. At each position a pigeon is launched from one side of the field and then the other. At certain positions, pigeons are launched from both sides of the field simultaneously. This is a true game of skill, simulating skeet shooting down to the last detail.

TRAPSHOOT allows one to five shotgunners to test their markmanship. The trap firing range has five positions where the one to five players shoot from. Each player is at a different location on the field. The challenge is to shoot pigeons out of the sky which launch at random trajectories. The challenge is to hit the pigeons while they are still in gun range.

SKEETSHOOT and TRAPSHOOT both allow you to control the size and speed of the pigeons and the width of your shotgun spray. Realistic sound-effects and HI-RES animation combine to make this simulation unparrelled for the APPLE.

DEALER INQUIRIES INVITED!

ALL SOFTWARE SHIPPED SAME DAY.
PHONE ORDER: (805) 522-8772

ON—LINE SYSTEMS, 772 NO HOLBROOK, SIMI, CA 93065

SEND TO: ______

		QUANTITY	TOTA
Hi-Res Adventure/Disk	$24.95		
Skeetshoot/Cassette	$14.95		
Trapshoot/Cassette	$ 9.95		
Skeetshoot/Disk	$19.95		
Trapshoot/Disk	$14.95		
Hi-Res Adventures & Skeetshoot/Disk	$37.50		
		Subtotal	
Payment: Check______		6% tax (Calif.)	
Master Chg/Visa #______		Shipping	$1.00
Expires:______		Total	

On-Line Systems' first advertisement, featuring Hi-Res Adventure ("Mystery House"). Published in *MICRO: The 6502 Journal*, May 1980, 80. From the collection of the author.

space in the back of the May 1980 issue of *MICRO: The 6502 Journal*, a hobbyist periodical for microcomputer enthusiasts working with the 6502 microprocessor.[36] *MICRO* ran black-and-white pages full of advertisements with low production value and minimal typesetting, making it a reasonable venue for an as yet barely existent company with a limited budget and no in-house layout artist. *MICRO* was also a magazine with a deeply technical readership, one that would immediately recognize the graphical innovation Ken Williams had coaxed out of the Apple's hi-res mode. To save on typesetting costs, Roberta designed the ad herself, cutting and pasting the words and images.[37]

In addition to splashing out on full-page advertising, Ken had rustled up a couple of other products to give their startup the appearance of larger scope. In their *MICRO* ad, the Williamses offered *Hi-Res Adventure ("Mystery House")* alongside two arcade-style games, *Skeetshoot* and *Trapshoot*, both programmed by an unnamed friend of Ken's.[38] *Mystery House* sold for $24.95—about $10 to $15 more than was typical for computer games—or could be bundled with *Skeetshoot* and *Trapshoot* for $37.50. The fate of *Skeetshoot* and *Trapshoot* is unknown. They existed to give the impression that there was more than a single game to sell but were quickly pulled from the Williamses' offerings following *Mystery House*'s success.[39]

The ad itself was a text-dense page describing the premise and wonders of an adventure game ("one who goes on an adventure is a venturer," the copy casually explains), composed in the unfocused, overwrought language that comes with having no market standard for how to sell microcomputer game software. The ad copy was inset by two photographs of *Mystery House*'s illustrations, rudimentary forms of what we would think of as a screenshot today.[40] Given that most game advertising used only hand-drawn illustrations, or no images at all, in marketing their products, the Williamses' advertisement made an implicitly dynamic claim for the visual appeal of their product. The advertisement also included a sales/shipping form customers were intended to cut out themselves, with instructions to make out checks out to "On-Line Systems," a company name that was a "holdover from Ken's vision of selling the respectable kind of business software for the Apple that he did in his consulting for online computer firms."[41] Orders by check, Master Charge, or Visa were received at "772 No Holbrook, Simi, CA, 93065"—the Williamses' home address.

Home production was a hallmark of many emerging microcomputer software companies, especially those that self-published.

Lacking a recognizable business model and uncertain of retailer and consumer demand, many software producers worked out of kitchens, garages, car trunks, and other makeshift spaces until they drew enough revenue and needed enough employees to warrant office space (or in some cases, until the companies attracted the notice of local zoning boards). The Williamses were no different: they soon found themselves selling *Mystery House* by the hundreds out of their home, fielding orders and giving hints to the game's puzzles on the family telephone line (805-522-8772).[42] By day, Ken would go off to a programming job while Roberta cared for house and children while packaging 5.25-inch game disks and simple photocopied documentation sheets into Ziploc bags and mailing out orders (see pl. 12).[43] As Steven Levy recounts in *Hackers*, "Ken and Roberta made eleven thousand dollars that May. In June, they made twenty thousand dollars. July was thirty thousand. Their Simi Valley house was becoming a money machine."[44] A more amusing measure might be Roberta Williams's recollection of wheeling a shopping cart full of Ziploc bags out of the grocery store.[45]

These steps were typical of independent software developers who published their own games and attempted to grow those efforts into a larger business. The Williamses may have lacked the refined marketing, packaging, and branding expertise that marked the release of *VisiCalc* just half a year earlier, but they weren't selling to the same market. Their challenge wasn't to convince people who weren't computer users to buy computers; *Mystery House*, like all games, lacked any appeal as a consumer necessity. Rather, the Williamses directed their marketing efforts to buyers who already owned or were ready to buy an Apple II and wanted to experience something novel with the system. To supplement their mail-order business, the Williamses also took *Mystery House* to local computer stores in the Los Angeles area to demonstrate the program for employees, something they already had experience with from running their own software distribution gig in early 1980 (Ken dropped the distribution enterprise once running On-Line Systems became a full-fledged side business).[46] According to Ken's recollection, retailers responded positively to the game, believing it helped them show off the Apple II to prospective buyers.[47] The program was compelling enough to be reviewed more than half a year later in the Christmas 1980 issue of *Creative Computing*, where it was praised for its "very nice" graphics, "showing the rooms and objects in detail."[48]

The kind of homegrown success story that *Mystery House* represents—just one of many examples in this nascent period of the computer game industry—would have been impossible in the arcade or console industry, where access to the development resources was more closely guarded. Thus, while it may be tempting to think of "games" as a single category across arcades, consoles, and microcomputers, the vastly different realities of production and consumption from platform to platform had practical consequences, as microcomputers spawned a thriving cottage industry for game development that did not exist for arcades or consoles. Game software, in its turn, helped people explore not only what a computer could do but also what they *wanted* a computer to do—a realm of democratized experimentation that arcades and consoles similarly lacked. This symbiotic relationship between microcomputers and their software would prove ideal soil for the burgeoning of a whole new industry.

* * *

In October 1980, the Apple II enthusiast magazine *Softalk* published its first ever best-seller list, ranking the top thirty Apple II software packages by sales.[49] The story was clear: *VisiCalc* may have been king, but its court was made of games. Of the twenty-nine software programs that ranked below *VisiCalc*, twenty-one were interactive entertainments of some fashion—including the Williamses' *Mystery House*, which was listed seventh overall.[50] At the top of the chart was the original *Flight Simulator*, one of the most enduringly popular games of the early Apple II era, followed by a motley mix of role-playing games (RPGs), adventure games, arcade copycats, computerized war games, and board game simulations: *Rescue at Rigel*, *Sargon II*, *Odyssey*, *Gammon Gambler*. A few would stay on that list for a year

The Top Thirty

1.	94.06	*VisiCalc*, Personal Software
2.	87.50	*Flight Simulator*, Sublogic
3.	79.06	*Bill Budge's Space Album*, California Pacific
4.	62.50	*Sargon II*, Hayden
5.	61.56	*Odyssey*, Synergistic Software
6.	60.62	*Adventure*, Microsoft
7.	56.25	*Hi-Res Adventure: Mystery House*, On-Line Systems
8.	51.25	*Typing Tutor*, Microsoft
9.	46.56	*Temple of Apshai*, Automated Simulations
10.	44.38	*Bill Budge's Trilogy*, California Pacific
11.	44.06	*Morloc's Tower*, Automated Simulations
12.	43.44	*Head On*, California Pacific
13.	43.13	*Rescue at Rigel*, Automated Simulations
14.	41.88	*Datestones of Ryn*, Automated Simulations
	41.88	*CCA Data Management System*, Personal Software
16.	40.31	*Super Invader*, Creative Computing Software
17.	37.81	*Wilderness Campaign*, Synergistic Software
18.	37.50	*Bill Budge's 3-D Graphics System*, California Pacific
19.	36.25	*Easy Writer*, Information Unlimited
20.	33.75	*Asteroids in Space*, Quality Software
21.	33.44	*Computer Bismarck*, Strategic Simulations
22.	31.88	*Apple Writer*, Apple Computer
	31.88	*Gammon Gambler*, Personal Software
24.	31.25	Scott Adams *Adventures*, Creative Computing Software and Adventure International
25.	29.69	*Computer Ambush*, Strategic Simulations
26.	28.44	*Tuesday Night Football*, Shoestring Software
27.	24.06	*Apple-Doc*, Southwestern Data Systems
	24.06	*Tranquility Base*, Stoneware
29.	22.50	*The Controller*, Apple Computer
	22.50	*Apple Plot*, Apple Computer

Softalk's "The Top Thirty" Apple II software best-sellers list, October 1980, 27. Image courtesy The Strong, Rochester, New York.

or more, but most were destined to be swept away in the rushing tide of new releases that followed every month. That was the funny thing about games: just how *many* of them there were. No one had use for owning half a dozen or more word processors, or inventory databases, or checkbook balancers, but games were defined by the fact that they were *consumed*, not *used*.[51] Most Apple II owners purchased at least a couple, and the true enthusiasts pored over floppy disks the way music aficionados curated record collections or book lovers coveted their shelves. Even if no individual game could beat *VisiCalc*, all of their sales together give different historical testimony about what people were *also* doing with their Apple IIs.

The late 1970s through the mid-1980s mark the beginning of a recognizable American computer game *industry*—meaning not just creating and circulating computer games for personal amusement, but leveraging such software as a commodity in the way the arcade and console industries had already been doing through much of the 1970s. Games may not have been the angle futurists and industry prognosticators wanted to offer a wary public on the benefits of the world's most expensive new consumer appliance, but games *sold*, and in great numbers, for every type of microcomputer on the market. Across all types of microcomputers, games were second only to business software in terms of sales revenue—roughly $540 million, or 34 percent of the $1.5 billion microcomputer software market circa 1983—and were generally estimated to be largest in sheer number of unit sales.[52] As one industry market report noted, even Osborne Computer, which specialized in portable microcomputers targeted to high-level executives, was releasing games for its system, proving "even the 'straightest' of business users play games on their machines."[53]

The first half decade of the computer game industry was a period of rapid transformation, with significant shifts in industry trends occurring roughly every sixteen to twenty-four months between 1979 and 1984. In its earliest incarnation, from late 1979 to mid-1981, companies were typically unpolished and undercapitalized, games were often buggy, and marketing was rudimentary. Yet with sparse competition, hits were fairly easy to come by, provided the software had some claim to legitimate novelty. Early movers in this space included On-Line Systems, SubLogic, California Pacific, Automated Simulations, Strategic Simulations, and Adventure International. In all cases these were ventures that, like On-Line Systems, had started as a lark by the founders, often with the intent of selling just a single product, but turned serious as they encountered an unexpectedly enthusiastic market. In these moments of haphazard success, company founders experienced the acute sensation of the so-called computing revolution—underwritten, of course, by the material financial gains they were suddenly reaping. Interpreting their good fortune as an indication of their own business acuity, many of these early movers quickly expanded their product lines to serve a market many people didn't even realize existed, formalizing themselves as publishers rather than simply self-publishing developers.

Softalk's October 1980 best-seller list gives a good sense of the wide range of play options made available by these early publishers. A few were action-themed space shooters and arcade knockoffs, works like *Asteroids in Space* and *Tranquility Base*, but many more illustrated the expansiveness of the microcomputer as a game-playing platform distinct from arcades and consoles. *Flight Simulator* had its moments of tension but was also a game of slow-going command-and-control, of reading dials and making careful choices. Adventure games like *Mystery House* and the Williamses' fall 1980 release, *Wizard and the Princess*, favored investigation and discovery, negotiated through language; these are games in which the primary struggle was a puzzle waiting to be solved. Strategy and role-playing games, like *Odyssey*, *Rescue at Rigel*, and *Temple of Apshai*, invited both reflex play and resource management, as players optimized exploration around the balancing act of health, wealth, weapons, and equipment. In the early creations of these greenhorn software developers, we see the emergence of unique interactive experiences that could not be supported by the economic logic of the arcade, which required short play duration and high challenge in order to induce coin drop, or the technical limitations of consoles, which reduced input to buttons, paddles, or joysticks.

Yet however innovative or novel many of these creations were, the microcomputer software industry was not yet at a stage of development where any of these publishers would have necessarily identified as *game* companies. Entertainment software might have been a low-stakes entry point for curious and idle programmers with an entrepreneurial streak, but the industry was still tenuous. Software publishers survived only if hardware manufacturers thrived, and expensive hardware like the Apple II sold only if you could prove it did things *besides* games.[54] These early company founders imagined themselves as ambassadors for the power of the microcomputer as an ambidextrous appliance and a programmable tool. The magic of the microcomputer was that it could be many things, and early publishers sought to prove this point by experimenting in other software categories, sometimes to great success. Ken Williams packaged the software he wrote to make On-Line's adventure games and sold it as a set of graphics utilities, and he released multiple word processors throughout the early 1980s. Brøderbund's *Bank Street Writer* would become one of the most popular home word processors of the era. Similarly, publishers of more "serious" software, such as Microsoft and Personal Software, also provided games in their portfolio. Thus while game development and publishing were fairly explicit among arcade and console software providers, these practices were less definitive in the microcomputer software industry, where games were always operating within a larger consumer software market.

The early, unpolished phase of Apple II game software development took on a new shape beginning around mid- to late 1981, as the relaxed pace and casual operations of many of these companies gave way to accelerated release cycles, rapid internal growth, and expanding product to other microcomputer systems. Joining the ranks of the earliest companies were new entrants, arriving almost monthly: Muse, Sirius Software, Penguin Software, and Brøderbund all emerged as industry leaders in their own right between 1981 and 1982. In the face of increased competition, companies rapidly built bandwidth for engaging in quality assurance, customer support, and marketing, lest buggy games garner negative reviews or homebrewed advertisements make products look unreliable.

The pace of releases also increased, as companies looked to increase their market presence and stave off competition. One strategy for dominating the market was to flood it, releasing games every few months in the hope that something became a hit—or if not, that subsequent releases would at least stabilize revenue overall even as each individual game quickly winked out. Arcade-style games proved especially quick to crank out. Approaching game production with a hobbyist-hacker ethic, developers frequently cribbed game mechanics from actual console or arcade games, resulting in unofficial Apple conversions or "borrowed" games.[55] These arcade copycats, known as clones today, were so common that *BYTE* ran an entire chart comparing the specs of eight different *Asteroid*-like games (seven of which were programmed for the Apple II).[56] Of the ten arcade games On-Line Systems placed on the *Softalk* best-seller list between 1980 and 1984, eight were clones of popular arcade games.[57]

Increased competition pushed publishers to exploit new revenue pathways, leading many publishers to begin releasing their games to other systems.[58] Sometimes called "porting," the process involved taking a game's original code base and reprogramming it to work on other systems, often those with same underlying hardware architecture. This explains the tendency for Apple II publishers to most frequently convert their games to the Commodore VIC-20, the Commodore 64, and the Atari 800, all of which were based on the same MOS 6502 microprocessor as the Apple.[59] Yet porting was time-intensive. Though publishers didn't have to develop new ideas, programmers still had to optimize a game-playing experience originated on the Apple II for systems with different hardware architectures. In some cases, companies hired programmers solely to do porting—thus keeping their best Apple II programmers focused on developing new products and letting others toil through the less glamorous work of code conversion.

Altogether, this excess of product began shortening the life cycle of a best seller to between three and six months. Many more games achieved notability for only a month or two and then fell away, or never trended at all. On-Line Systems proves illustrative: despite early long-standing best sellers such as *Mystery House* and *Wizard and the Princess*, sustained popularity proved increasingly elusive for many of the Williamses' products. Most of their subsequent adventure games sank quickly into anonymity, while a couple of moderately well-performing arcade games and a successful word processor,

Screenwriter II, survived on the Apple II best-seller lists. Computer game developers were starting to feel the pinch of competition, and even more was on the horizon.

* * *

By 1982, rapid economic growth in the microcomputer game software sector was drawing attention from several corners, even as the market grew ever more crowded. Venture capitalists, alongside new entrants into software publishing such as book publishers and other media giants, significantly increased the financial scale of the entertainment software market. Furthermore, changes in the hardware scene—namely, the release of numerous convergent systems that functioned as hybrids of microcomputers and video game consoles—encouraged entertainment software publishers to divide their attention among many competing systems, as well as several media formats. Ultimately, a massive proliferation in game software products, all being hotly underwritten by investor fever dreams based on delirious projections of market growth, would prime the industry for internal collapse.

While venture capitalists were not involved in the earliest phases of home entertainment software development and publishing, several were quick to turn their attention to the industry once it began demonstrating rapid financial growth, around 1981.[60] The finance industry was acutely aware of the remarkable success of Apple Computer itself, which had already garnered phenomenal profit for its earliest investors, making it a darling of the business and technology press. Similarly, Atari seemed indomitable in the early 1980s, widely recognized as one of the fastest-growing companies in American history, raking in nearly $290 million in operating income in 1981 alone and dwarfing its parent company, Warner Communications.[61]

Hoping to score their own big wins, venture capitalists began exploring investment opportunities with top entertainment software providers such as On-Line Systems, Brøderbund, and Automated Simulations, as well as fresh startups like Electronic Arts, all during the early 1980s.[62] Venture capital in the early US software industry worked the same way it did for companies like Apple Computer: venture capital firms offered large influxes of cash in exchange for a significant percentage of stock in an early-stage company, and often seats on the board. Such financial dealings were "ventures" in the sense that they were highly speculative; most companies would never make good on the hearty valuations offered by their venture capital investors. But the companies that did achieve expectations—either by going public or being acquired—inevitably did so at such a vast scale that they more than compensated for the high degree of failure or breakeven in the venture capital investment pool overall.[63] On-Line Systems was one of the first to jump into venture capital, taking a cash infusion of $1.2 million in 1982 from the East Coast venture capital firm TA Associates in exchange for 24 percent of the company.[64] While the deal was significant, it was perceived to match On-Line Systems' respective value as the largest independent software firm in the country, with 11 percent market share across all consumer software.[65]

This influx of venture capital also brought a rebranding: On-Line Systems would take the name Sierra On-Line, an homage to the company's proximity to Yosemite National Park in the heart of the Sierra Nevada. Blending technical terminology with one of the nation's natural wonders, the new name and logo was intended to extend the company's brand appeal beyond tech-friendly early adopters to future consumers who might feel wary about the role of computer technology in their lives. And of course, for all these companies, venture capital brought new stakes—and stakeholders—to the table. Anecdotal and journalistic accounts of this heady, gold rush moment are rife with stories of increasing administrative bureaucracy, clashes between management and developers, increased marketing budgets, and newly internalized pressures to professionalize and mature industrial software development, manufacturing, and distribution practices.[66]

Following on the heels of the venture capitalists were established media corporations seeking their own footholds in microcomputer entertainment. This was particularly common among book publishers and broadcast media giants, which felt pressured to catch up to the abrupt arrival of the much-touted computer revolution. In fall 1982, CBS Software, a subsidiary of CBS, one of America's oldest television broadcast companies, formed a software unit, on the premise that CBS was "in the business of delivering a message into the home. . . . The home computer [is] just another way of delivering that message."[67] Similarly, in 1983, the media conglomerate Cox Enterprises purchased Creative Software, "looking ahead to a day when it may circulate its newspapers through computers."[68] Other popular media brands, including Reader's Digest and Scholastic, also entered the software market. Doug Carlston, the founder of Brøderbund, saw this as an industry transition in which software publishers were imagined as a kind of marketing exercise for larger conglomerates: "the company name is the focus of attention, not the individual products."[69]

As Carlston's observation suggests, the goal of these newcomer companies was not necessarily to threaten the entertainment software industry directly. Indeed, many of them avoided making any explicit inroads in the games market, still operating under the somewhat naive assumption that games were toys for consoles while software was serious business. Instead, companies flooded the market with family-friendly "educational" titles rather than outright computer games (see chapter 7 for more discussion of the education software market). How this affected the game software publishers, however, is the extent to which they imagined "education" as falling under the purview of "entertainment"—a line that could get quite blurry, given that educational titles were often merely learning exercises skinned around more traditional game mechanics.

The larger impact the arrival of these major media players had on the entertainment software sector, however, was the extent to which they pushed startup software publishers like Sierra On-Line toward more serious levels of competition. Flush with newfound credibility and eager to maintain brand recognition, Sierra On-Line became notable for pursuing entertainment partnerships to make games with intellectual property (IP) ranging from Jim Henson's *The Dark Crystal* to the comics *The Wizard of Id* and *B.C.* to characters from Disney films (licensed for a line of children's educational games).[70] These kinds of licensing arrangements were expensive and could only be secured with the investment capital being supplied by the finance sector. While many journalists at the time covered these deals as proof that computer software was arriving as a serious cultural form, they retrospectively take on the character of a forced growth pattern stimulated not by consumer demand but by anxiety over market position.

The final significant impact on the Apple II entertainment software industry during this time was the proliferation of microcomputing platforms, in particular, the arrival of a new class of computer technologies that were hybrids of consoles and microcomputers. Since the release of the earliest US at-home game consoles, like the Magnavox Odyssey and the Atari *Pong* machine, consoles and microcomputers had operated as distinguishable technologies, even if they both afforded opportunities to play game software. That would change in 1979, when Atari released a pair of microcomputers based on the 6502 microprocessor, aimed at the small business and home computing markets, respectively: the $1,000 Atari 800 and its feature-stripped sibling, the $550 Atari 400. These two systems were both substantial microcomputers, outfitted with BASIC programming functionality, sound capabilities, and sophisticated graphical hardware design, but they also had the significant additional feature of a cartridge slot, allowing them to run software just like Atari's VCS console (though software was not compatible between Atari's line of microcomputers and the VCS). Cartridge-based software allowed users to skip the command line and immediately load the software, so nontechnical users could avoid anything that even resembled programming or textual communication with the machine. Hoping to attract "a consumer that had not necessarily had any prior computer experience," Atari's 400/800 line was an effort intended to increase market share—and counter the microcomputer's encroachment into the video game marketplace—by capitalizing on preexisting consumer familiarity with the operation of consumer electronics.[71] In the coming years, other hardware manufacturers from both sides of the aisle followed suit, producing a flurry of machines that blended a microcomputer and a video game console: the Commodore VIC-20 and Commodore 64, the Coleco ADAM, Texas Instruments' 99/4A, and the TRS-80 Color Computer (better known as the "CoCo"). Reporting from the floor of the winter 1983 Consumer Electronics Show, Russell Sipe wrote for *Computer Gaming World* that "the line of demarcation between personal computers and home videogame systems has disappeared."[72]

This influx of hybrid systems not only accelerated the practice of porting but also pushed computer game software publishers to see themselves in more direct competition with traditional video game manufacturers, especially given the video game industry's strategic lockdown on arcade IP. Before the arrival of hybrid systems, Apple II game software developers and publishers had grown successful making clones of popular arcade games, avoiding significant legal trouble by making the games' graphics look different (even if the games were played the same). But if cartridge manufacturers could easily produce their arcade-based IP for cartridge-loading microcomputers, there was simply no way a microcomputer publisher could compete with a game console maker's economies of scale—which meant ownership of IP became an important commodity in a way it had not previously been for microcomputer game software developers. In 1982, Ken Williams secured the magnetic media rights to Sega's arcade hit *Frogger*, ensuring that his company alone could release the game for any system that could accept floppy disks or cassettes (Parker Brothers paid for the cartridge media rights).[73] It would be the first time an arcade-based IP found official release on the Apple II, and it proved phenomenally popular for some microcomputer platforms, such as the Atari 800. Tens of thousands of units were ordered. As a Sierra On-Line sales rep recalled of this time, "All you could see was a tomorrow that was endless."[74]

In the span of three or four years, many of the early risers in the Apple II game software scene experienced a rapid ascent in an industry the likes of which had never existed. Many of these company founders had started off as friends, an underdog bond driven by the fact that they had all thought they were longshots. Yet over time, as the money was counted, as company shares were parceled out, as new competitors grew like a hydra's heads, as the industry's internal whirlwind made it all feel like an exhilarating competition for developers and publishers to one-up each other (all while padding the profits for investors and declaring it their privilege to do so), the wheels began coming off the wagon.

* * *

In history's rearview mirror, it certainly seems like people knew a bubble was forming. By 1983, the editorial pages of computer magazines were mottled with speculative wonderings. Robert Lock, editor in chief of *COMPUTE!*, wrote of a "crisis of expectations" as "the industry is simply not expanding at the rate many expected."[75] *Softalk*'s August 1983 coverage of that year's Summer Consumer Electronics Show reported a "software glut"—even as the industry's earliest entrepreneurs were "tak[ing] their place in the mainstream marketplace alongside firms like Mattel, Parker Brothers, CBS, FOX, Coleco, Imagic, Activision, Atari, and others."[76] "Industry analysts all concur that a 'shake out' in the personal computer arena is imminent," opined Maggie Canon, editor of the newly anointed Apple consumer magazine, *A+*, in her November 1983 launch issue. Mitch Kapor, developer of *Lotus 1-2-3*, described the impending disaster thus: "Competition is forcing a shakeout, yes; but companies are not going to explode in flames like the Hindenburg, or go under like the Titanic. They're more like the ships of Magellan's fleet. Some will make it around the globe; others will lose their way and sink almost without a sound." While Kapor effortlessly naturalizes the metaphor of Magellan's global escapades, it is also true that the ships were only lost because colonial enterprise had induced them to circumnavigate the globe to begin with. Software sales were flattening at the very moment everyone, from venture capitalists to book publishers, was trying to get into a business they had neither helped create nor adequately understood, yet were anxious not to miss out on.

The "Software Shakeout," as it was generally referred to in the tech press, was not unique to the computer game business, or to Apple. It was a confounding set of affairs: microcomputer software sales were doubling year over year, but those successes were generally held by a small number of incumbent hardware manufacturers, such as IBM, Tandy, Commodore, Apple, Atari, and Texas Instruments. Those six companies alone accounted for nearly 40 percent of software sales. Yet the microcomputer industry comprised more than five thousand software companies, the vast majority of which were scrabbling over the bottom 28 percent of the market. For small and even midsized companies, these were unsustainable dynamics, as the industry consolidated around well-resourced hardware manufacturers and a small number of independent software publishers. Acquisitions became rampant as smaller companies looked to get out of the market.[77]

These tensions would prove particularly acute within the game sector, where easy market entry in 1980–81 had led to a proliferation of small companies. Yet many game software publishers continued unabated, despite the fact that 1983 was the same year that a recession in the US video game market—known today as the North American video game crash—sank many video game publishers and even capsized Atari, resulting in flattening growth and warehouses full of unsellable product.[78] The assumption of technological progress so well internalized in the microcomputer industry blinkered many game publishers to the potential fallout. As Brøderbund cofounder, Doug Carlston, wrote of this period:

> The computer software people watched all the carnage in the video game industry with considerable complacency. Nineteen-eighty-three turned out to be a banner year for most of us. . . . These video game-based businesses were still seen only as software dinosaurs, mired in a market without a future, turning out trivial products for increasingly sophisticated consumers.

The reality of their situation would not begin to set in until 1984. As one industry market report described, Ken Williams was dismayed to discover that Sierra On-Line's software sales were not growing at the same rate as hardware sales, resulting in a failure to meet the financial expectations placed on his company by heady investors.[79] The market was beyond saturated, and the growth anticipated by the arrival of the hybrid machines, and the new consumers they might drag in their wake, never came. Companies like Sierra On-Line had dumped massive upfront capital into securing the production of ROM cartridges for machines like the Commodore VIC-20, the Atari 800, and the Commodore 64 while also porting their floppy disk products across this proliferation of new platforms. This meant that for a single game, they now had to program and manufacture a variety of ROM cartridge and floppy disk formats—increasing the costs of production even as sales were plateauing. Compounding the issue, cartridges were more expensive to produce than floppy disks, and unlike floppy disks, they weren't reusable if a game sold poorly and stock was returned to the publisher. From development to manufacturing to distribution, costs were mounting as revenue was narrowing.

The bubble didn't pop so much as it created a rolling blackout. For all the copies of *VisiCalc* surely lying around the desks of industry founders, it would take quite a bit for companies like Sierra On-Line to realize that while growth was increasing, the rate of growth was slowing. Entire swaths of the industry cratered and never made a comeback. Major companies hawked what remained of their warehouses at fractions of their value. Magazines folded, disappearing overnight. A whole generation of early stars were never heard from again. Sierra On-Line would lay off 100 of its 130 workers. Yet despite the damage done by overprojection and overcapitalization, the fallout was quickly normalized within the computer game industry. The lost companies became like Kapor's lost ships: accepted with ex post facto rationalizations that this was the price of doing business, the proof that computer games were a *real* industry, without any questioning of the relentless quest for profit or lack of foresight that had sunk them.

Sierra On-Line survived the Software Shakeout, going on to see another decade of successful releases. Even this success didn't last, however. The Williamses sold the company in 1996, after which an accounting scandal resulted in the brand being passed from one buyer to another, losing employees and market share each time. Eventually, the business was shuttered altogether—the company ultimately proving as ephemeral as the games it created.

Yet what would prove most durable about this early period of computer game history was not specific companies or games but the new realities of production and consumption that it brought into being—realities that continue to shape the relationship between people and their computers today. In the realm of production, this period saw the congealing of an industry around certain practices, from divisions of labor and genres of content to expectations of play and profit, that persist into the present. In the realm of consumption, this was the moment computers left the office and entered the home, becoming integral not only to people's labor but also—through computer games—to their leisure.

Ultimately, it was the Apple II that benefited most from this enthusiastic community of cultural producers who got swept along in entrepreneurial fantasies. For every dozen or so derivative arcade clones that swarmed the market, there was a game or two that pushed new edges and drove excitement about the platform: *Flight Simulator*, *Mystery House*, *Loderunner*, and others. Though often treated as a frivolous category of software production, "ephemeral in the extreme," as the software historian Martin Campbell-Kelly has written, computer games were a vital component of the larger economic, technological, and cultural ecosystem that developed around the prospect of personal computing, and the Apple II in particular, given its graphical capabilities.[80] Even as consumers might tie their purchasing rationalizations to the utility of spreadsheet software or the benefit of readying their children for the Information Age, games made computers legible as technologies in ways that many other kinds of software did not. Games underscored what the game programmer Chris Crawford termed, in 1984, the fundamental "flexibility" and "plasticity" of computing: the power computers had to change the rules of a system or alter interactivity through the immediacy of software programming.[81]

Like arcade and console games, games designed for microcomputers would become software for the exploration of new modes of sensation, attention, and immersion, and they could do so through a broader range of genres than the dedicated hardware of arcades and consoles. Because of this capacity, games seemingly exemplified what the techno-futurist Ted Nelson prophesied, in 1977, as the "new kind of mental life" available to us through computers, whereby through "abstracting ideas and transforming them into vivid experience, they expand our minds."[82] Whether or not such prognostications proved true was less important, historically speaking, than whether such ideas cultivated a culture of anticipation about what games could make possible for human experience—even as these explorations quickly fell into familiar patterns of capital accumulation and consumerism. Games may have lived free from the demands of productivity that fell on other types of software, but as far as economics was concerned, everything was business as usual.

5

Utilities

Locksmith

Robert Tripp was starting out March 1981 with a mea culpa. As the editor and publisher of *MICRO: The 6502 Journal*, Tripp was in the unenviable position of having to pen an apology for running an ad for the disk copy utility program *Locksmith*.[1] He started out by insisting on the obvious, that *MICRO* was "unconditionally opposed to the illegal copying of software listings, cassettes, diskettes or any other protected material."[2] He likened the copying of software to the photocopying of a magazine and acknowledged that *MICRO* would have no livelihood if readers could simply get the content free or at minimal cost. He then went about surveying the "hidden costs" of illegally duplicated software, or "copywrongs," as he put it, in a fizzled attempt at humor. According to Tripp, software piracy increased costs for consumers, added technical headaches, and, in the very long term, deprived developers of the royalties necessary to sustain themselves and their programming practices. "The only person who benefits from 'copywrong' is the thief," Tripp wrote. "Everyone else loses in the long run."

The initial *Locksmith* advertisement. Published in *MICRO: The 6502 Journal*, January 1981, 80. Image from the collection of Jason Scott.

Tripp's editorial was motivated by something that had never happened in the world of microcomputer journalism before: the threat of an advertiser boycott. The software publishers who formed a sizable contingent of Tripp's advertising base were incensed over the *Locksmith* ad, which they believed served to enable software piracy. Released by the newly formed Omega Software Systems, an otherwise anonymous publisher with no prior products to its name, *Locksmith* promised to copy the "uncopyable." In other words, it allowed users to make duplications of floppy disks that were otherwise locked by the varied copy protection schemes used by publishers as a loss prevention mechanism to stanch piracy.

Understanding why a group of software publishers might threaten to ruin a niche microcomputer hobbyist magazine over another company's ad requires understanding the tangled net of industrial tensions that emerged between the producers and the consumers of microcomputer software in the early 1980s. As a consumer microcomputing market began to flourish, developers became alert to the risks software piracy posed to their burgeoning industry. If no one paid for software, they worried, who would bother to write it, and how would the industry grow? Thus began the drama of *copy protection*, an industrial loss prevention practice wherein companies used a combination of hardware and software techniques to scramble the data on software media formats, typically 5.25-inch floppy disks, so that copying the disk was no longer possible by conventional means. While the goal of this subtle bit of friction was to throttle piracy, it also prevented users from creating backup copies of software they legally owned, or otherwise accessing the code itself.

Copy protection centralizes unique tensions around the status of software in the early 1980s, particularly with regard to its ownership. While these questions had not mattered much in laissez-faire research settings or in corporations that had no incentive not to play by the book and pay the necessary license fees, the consumer consumption of software and publishers' efforts to control that consumption were subject to furious cultural, economic, technical, and legal debate. Was software a good or a service? Did users have a right to access the code itself or only its end result? Was preventing users from making backup copies a form of industrial overreach or even consumer abuse? No computer enthusiast magazine of the period, and no software publisher or consumer, went untouched by the roiling debate over copy protection.

There are no definitive answers to such questions, but putting a historical spin on the issue offers a richer query: Why were such questions ever concerns to begin with? As it turns out, the way people used their computers matters a great deal for understanding why copy protection (and copy breaking) became such a fiercely contested issue. Copy protection could interfere with personal computing's most mundane operations—the quotidian ability to *use* your software. Taking seriously the claims of those who supported *Locksmith*, not to enhance piracy operations, but to make their own interactions with software more efficient, secured, or productive, this chapter deploys *Locksmith* as a historical artifact to productively center the habits and practices (and challenges and frustrations) of everyday computing. *Locksmith*'s very existence bookmarks a convergence of technical affordances, economic incentives, and social practices unique to microcomputer software ownership in the late 1970s through the mid-1980s.

Within the larger ecology of Apple II software, *Locksmith* would be categorized as a *utility*. Utilities functioned as a categorical grab bag for products that were tools for the Apple II, commonly united by their deployment as a technical means toward typically technical ends. Decked with beefy, quasi-explanatory names like *DOS Boss*, *Super Kram*, *Quickloader*, and *Bag of Tricks*, utilities were programs for making disk load times quicker, directory searches faster, coding more legible, graphics programming less complicated, deleted files recoverable, and yes—copy protection breakable. This was software that framed the relationship to the computer not as a product to consume but as a boundary to negotiate. As a category, utilities were largely the purview of hobbyists, hackers, and computing professionals—the types of users who are most vividly remembered in the historical imagination, although they became increasingly a niche as the microcomputer went mainstream. Yet as *Locksmith* proves, not all utilities were beyond the reach of a nontechnical user.

Locksmith is not a reflection of all utilities writ large but rather an anomaly that nonetheless intricately reveals its sociotechnical surround. Tracing the debate over disk duplication that provoked the program's very existence centralizes one of the great contestations of this period: usability versus profitability. The story of *Locksmith* helps contextualize the material practice of software piracy not as a dilemma about whether or not "information wants to be free" but as a case study on the contesting norms of different users and their expectations of computer productivity.

This approach is necessary—and valuable—for another reason: *Locksmith* is unlike any of the other pieces of software discussed in this book. It is not well remembered, nor was it a dramatic technical innovation. Its advertisements ran for only a couple of months before being pulled by nearly every enthusiast magazine on the market. Little is known about its production history, its developer, or its publisher. In this, *Locksmith* represents a silent majority of software from this period. Unlike *VisiCalc* or *Mystery House*, most early computer software is not well represented in the historical record, not documented by an abundance of material that is simply waiting to be dusted off and read. In reality, lack of documentation is the rule for the vast majority of software products made during the late 1970s and into the mid-1980s. This is especially true for utility software, which typically did not draw commanding profits or offer readily demonstrable or interactive experiences. In other words, if *VisiCalc* is the defining landmass of American software history, the coastline is dotted with thousands, tens of thousands, of barren archipelagos representing all that may ever be known about other software from this period. In this way, then, *Locksmith* reflects a truer set of historical affairs than any of the other software covered herein: it is, and will likely remain, lost to time. Only by following the software's afterimages can we begin to pull together an impression of how, and in what ways, *Locksmith* may have mattered.

* * *

Think of utilities as software but meta: software for and about using the computer. Rarely an end in themselves, utilities circulated as tools for reducing friction, streamlining redundant tasks, troubleshooting, or fixing user errors. The meaning of the term shifted as computing crossed from the professional and research contexts ruled by mainframes and minicomputers to the more idiosyncratic and personalized spaces common to microcomputers. In the minicomputer era, for example, "utilities" meant software that assisted in systems management—a category that was distinct from programming languages, maintenance software, or tools for debugging, disassembly, and dumping.[3] In microcomputing contexts, however, "utilities" became an umbrella term for any software that pertained to the operation and maintenance of the computer itself or the production of software. Given that highly technical users were only a narrow subset of the microcomputing audience, magazines and retailers had little reason to make fine distinctions among such software. From an industry categorization perspective, programming languages, compilers, interpreters, assemblers, and even disk operating systems made up the same category as graphics applications, disk duplicators, indexers, and code editing tools.

Although some of the first programs developed for microcomputers were utilities, this type of software was commercialized more slowly than more self-explanatory applications like games, business software, or programs for calculating scientific functions. This is because in the earliest years of microcomputing, particularly in the mid- to late 1970s, microcomputers were still the domain of a professional technical class linked by robust hobbyist communities—skilled enough to either program their own utilities or make use of those shared by their peers. Consequently, programs that would later be classified as commercial utilities were routinely shared as code listings in hobbyist magazines or circulated among user groups, typically at little if any cost to the user. Six of the first fourteen programs included in the Altair Users Group Software Library, as published in July 1975, were utilities: two assemblers, designed to convert program code into machine code legible to the Altair; a diagnostic program to search for memory access errors; a 7-byte program for clearing the Altair's memory; a debugging routine; and a program that could move another program from one memory location to another.[4] In all instances, these are programs in service of other programmatic goals. Submitted to the library "by interested users who desire to provide a service to other users," these programs were available to members of the Altair Users Group for merely the cost of duplication and shipping (typically about $5).[5]

Hobbyist magazines were similarly full of printed listings providing useful tools for dedicated, routine microcomputer enthusiasts. A good example appears in the November–December 1978 issue of *Creative Computing*: a program called INDXA that produces file indexes for data storage cassettes.[6] This listing, like many from the period, is accompanied by an article describing a personal computing situation that was so frustrating it motivated the author, Rod Hallen, to create custom software to resolve it.

> I'm showing off my computer to a friend. After a few minutes of rolling the dice, I decide to run my electronic slot machine for him. Now where is it? I know it's on one of these tapes. I think it's this one. Out goes the old tape and in goes the new. Load. Run. No! That's my checkbook balancer program. Wrong tape!
>
> Sound familiar? The accessibility of my programs dropped as a direct result of my increasing tape collection. I needed some way to keep track of my various tapes. My first step was a hand-written loose-leaf catalog. Even though it was primitive by computer standards, I at least knew where everything was. But there had to be a better way.[7]

The reason Hallen had so many tapes was because he was using one cassette tape per program, which was the norm for hobbyist computer users at the time. So part of Hallen's "better way" was to devise a software utility that would allow him to save multiple software programs onto a single cassette tape. To do so, his utility needed to track where on the tape each program began and to keep an index of that information to allow users to choose from a menu of programs on a given tape.[8]

For Hallen, the point of creating such a utility was to improve the overall experience of using his computer; his frustrations were contextual, specific to his experiences as a user, a response to his own fumbling and forgetting. While his initial problem of not knowing what was held on what tape could have been solved by labeling the tapes themselves, that solution would not have addressed the inefficiency of having many programs spread out across many individual tapes. Indexing in this manner allowed Hallen to organize his software collection by type, creating cassettes dedicated to specific functions, such as business applications, mathematical routines, or games. As he writes, "The initial creation of all these files, catalogs, and indexes requires a certain amount of drudgery but, once they are on tape, your personal computing will be simpler, easier, and much more enjoyable."[9] While this listing is particularly illustrative of the kind of work utilities did for hobbyists, countless examples can be found in hobbyist magazines of the period.

While many utilities circulated in the DIY, shareware ecology of hobbyist computing, the commercialization of this kind of software follows trends consistent with the broader Apple II software market, in which amateur entrepreneurs released products with low-cost, homespun marketing and packaging and then leveraged early successes into opportunities to grow their operations and make the products more professional. The market for such products wasn't as vast as the market for business applications or computer games, but the heavy use and essential nature of this software made it ideal for modest commercial endeavors. Programming was challenging, no matter how devoted one was to the practice, and programmers were eager for something that might alleviate the tedium of developing software.

Two early successes in this market are *Apple-Doc* (1979) by Roger Wagner, founder of Southwestern Data Systems, and *Bill Budge's 3-D Graphics System and Game Tool* (1980), published by California Pacific Computer Company. Both programs landed on *Softalk*'s first best-seller list in October 1980, with more copies sold than several popular games and even some programs from Apple Computer. They also reflect essential qualities of the utilities category: most prominently, an intended audience of technically advanced users and the characterization of the software as a tool for supporting other kinds of programming work. For example, *Apple-Doc* described itself as "an aid to the development and documentation of Applesoft programs," allowing users to annotate their code, list or replace specific variables, and generally make software editing in Applesoft BASIC less tiresome for professional and hobbyist programmers alike.[10] Similarly, *Bill Budge's 3-D Graphics System* was designed to assist in the creation of displays and animations for 2D and 3D computer games written in BASIC, especially for "Apple users who don't know assembly language."[11] Despite the manual's insistence on ease of use, however, the software required the user to possess substantial programming skill, including the capacity to code the operations of the game itself in BASIC, in order for the utility to be of value. Furthermore,

Plate 1. Apple II microcomputer with a third-party monitor and a Disk II 5.25-inch floppy drive. Apple II released in 1977; Disk II first manufactured in 1978. Image courtesy the Division of Medicine and Science, National Museum of American History, Smithsonian Institution.

Plate 2. Scenes from *Desk Set* (top, 1957) and *Colossus: The Forbin Project* (bottom, 1970), illustrating representations of mainframe computers in midcentury popular culture. Screen captures by author.

Plate 3. Cover of the *PDP11 Processor Handbook* (1975), featuring a PDP-11 minicomputer with the desktop input/output terminal, which was manufactured by Digital Equipment Corporation beginning in 1970. The smaller scale of the PDP-11 illustrates the increasing miniaturization of computing systems over the course of the twentieth century. Image courtesy the Hewlett-Packard Company.

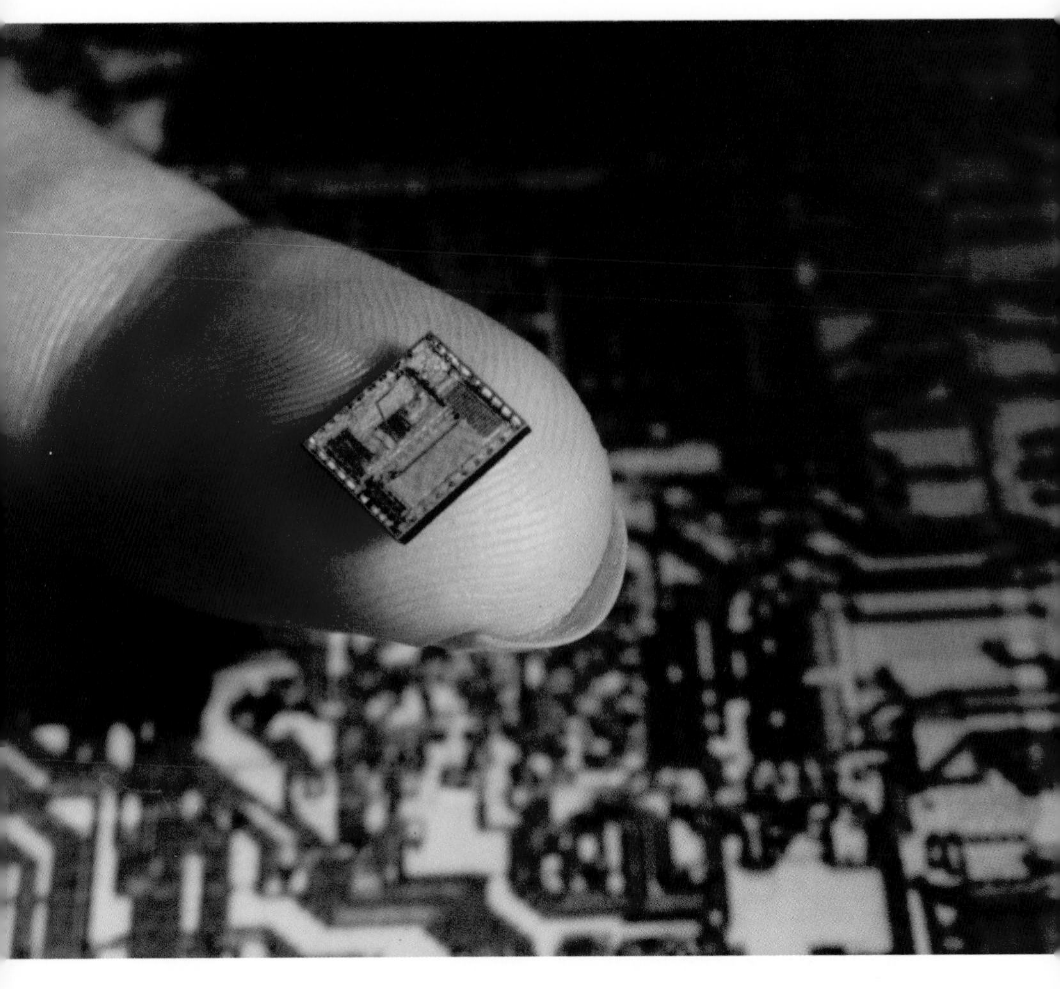

Plate 4. An Intel EPROM 8748 microprocessor chip, so small it can be held on the tip of a finger. Date unknown. Courtesy of the Computer History Museum.

HOW TO "READ" FM TUNER SPECIFICATIONS

Popular Electronics

WORLD'S LARGEST-SELLING ELECTRONICS MAGAZINE JANUARY 1975/75¢

PROJECT BREAKTHROUGH!

World's First Minicomputer Kit to Rival Commercial Models...

"ALTAIR 8800" SAVE OVER $1000

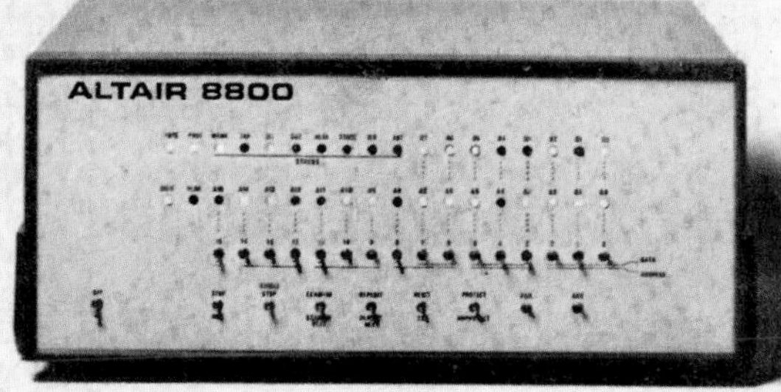

ALSO IN THIS ISSUE:

- An Under-$90 Scientific Calculator Project
- CCD's—TV Camera Tube Successor?
- Thyristor-Controlled Photoflashers

TEST REPORTS:

Technics 200 Speaker System
Pioneer RT-1011 Open-Reel Recorder
Tram Diamond-
Edmund Scienti
Hewlett-Packar

18101

Plate 5. Cover of *Popular Electronics*, January 1975, showcasing the release of the Altair 8800, the first widely advertised microcomputer intended for individual purchase, promoted at the time as a "minicomputer kit." Image courtesy www.worldradiohistory.com.

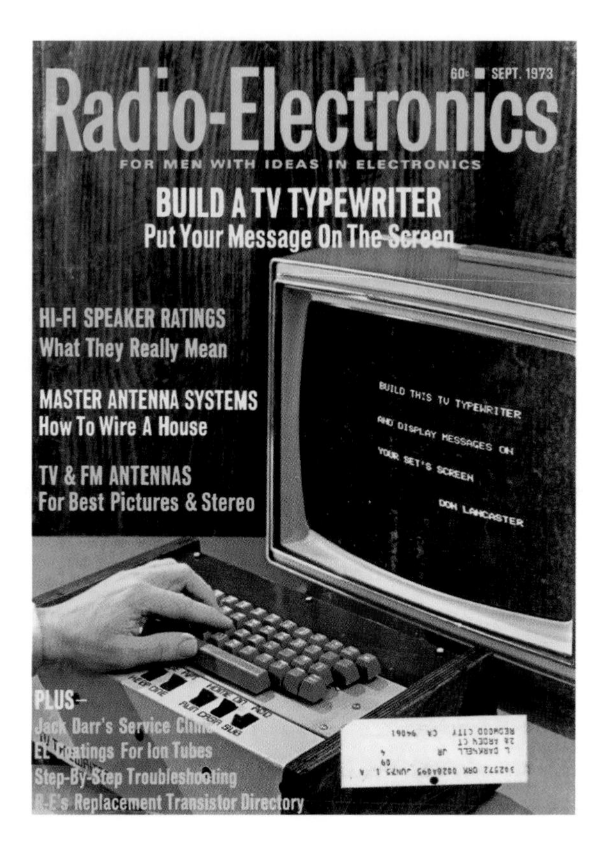

Plate 6. Cover of *Radio-Electronics*, September 1973, illustrating a popular mid-1970s DIY TV terminal project, Don Lancaster's "TV Typewriter." Image courtesy www.worldradiohistory.com.

Plate 7. Apple 1 boards packed in white boxes, stacked in a bedroom in Jobs's family home, 1976. Note the soldering iron and reel of solder on the desk, likely used to solder the chips to the board, and a television set for testing the completed boards. It is not clear if the boards in the boxes are "finished" Apple 1s or were waiting to be soldered. Image courtesy Steve Wozniak.

Plate 8. An Apple II with its lift-off cover removed. This feature makes the Apple II's central board immediately accessible to users. At the back of the Apple II are the system's eight expansion slots, allowing for the easy addition of peripherals or other forms of expanded functionality. In the image, three slots have boards plugged in, while the other five are empty. Photography by Tony Avelar, image courtesy Bloomberg via Getty Images.

Plate 9. The software and documentation for *VisiCalc*. The product's brown vinyl three ring binder evokes the tools and materials of a businessman rather than a microcomputer hobbyist. Image courtesy the Division of Medicine and Science, National Museum of American History, Smithsonian Institution.

Solve your personal energy crisis. Let VisiCalc™ Power do the work.

With a calculator, pencil and paper you can spend hours planning, projecting, writing, estimating, calculating, revising, erasing and recalculating as you work toward a decision.

Or with VisiCalc and your Apple* II you can explore many more options with a fraction of the time and effort you've spent before.

VisiCalc is a new breed of problem-solving software. Unlike prepackaged software that forces you into a computerized straight jacket, VisiCalc adapts itself to any numerical problem you have. You enter numbers, alphabetic titles and formulas on your keyboard. VisiCalc organizes and displays this information on the screen. You don't have to spend your time programming.

Your energy is better spent using the results than getting them.

Say you're a business manager and want to project your annual sales. Using the calculator, pencil and paper method, you'd lay out 12 months across a sheet and fill in lines and columns of figures on products, outlets, salespeople, etc. You'd calculate by hand the subtotals and summary figures. Then you'd start revising, erasing and recalculating. With VisiCalc, you simply fill in the same figures on an electronic "sheet of paper" and let the computer do the work.

Once your first projection is complete, you're ready to use VisiCalc's unique, powerful recalculation feature. It lets you ask "What if?", examining new options and planning for contingencies. "What if" sales drop 20 percent in March? Just type in the sales figure. VisiCalc instantly updates all other figures affected by March sales.

Or say you're an engineer working on a design problem and are wondering "What if that oscillation were damped by another 10 percent?" Or you're working on your family's expenses and wonder "What will happen to our entertainment budget if the heating bill goes up 15 percent this winter?" VisiCalc responds instantly to show you all the consequences of any change.

Once you see VisiCalc in action, you'll think of many more uses for its power. Ask your dealer for a demonstration and discover how VisiCalc can help you in your professional work and personal life.

You might find that VisiCalc alone is reason enough to own a personal computer.

VisiCalc is available now for Apple II computers, with versions for other personal computers coming soon. The Apple II version costs just $99.50 and requires a 32k disk system.

For the name and address of your nearest VisiCalc dealer, call (408) 745-7841 or write to Personal Software, Inc., Dept. B, 592 Weddell Dr., Sunnyvale, CA 94086. If your favorite dealer doesn't already carry Personal Software products, ask him to give us a call.

PERSONAL SOFTWARE

TM—VisiCalc is a trademark of Personal Software, Inc.
*Apple is a registered trademark of Apple Computer, Inc.

Circle 301 on inquiry card.

Plate 10. The first *VisiCalc* advertisement, designed by the ad firm Regis McKenna and published in *Byte*, September 1979. Image from the collection of Jason Scott.

Plate 11. Ken and Roberta Williams. Image published in *Softalk*, February 1981, 4. Photo by Brian Wilkinson. Image courtesy the Computer History Museum.

Plate 12. Plastic baggie packaging for *Mystery House*, ca. 1980. Image courtesy of Brad Herbert, SierraMuseum.com.

Plate 13. Apple Computer's launch campaign for the Apple II microcomputer. Published in *Byte*, June 1977, 14. From the collection of Jason Scott.

You've just run out of excuses for not owning a personal computer.

Clear the kitchen table. Bring in the color TV. Plug in your new Apple II,* and connect any standard cassette recorder/player. Now you're ready for an evening of discovery in the new world of personal computers.

Only Apple II makes it that easy. It's a complete, ready to use computer, not a kit. At $1298, it includes video graphics in 15 colors. It includes 8K bytes ROM and 4K bytes RAM—easily expandable to 48K bytes using 16K RAMs (see box). But you don't even need to know a RAM from a ROM to use and enjoy Apple II. For example, it's the first personal computer with a fast version of BASIC permanently stored in ROM. That means you can begin writing your own programs the first evening, even if you've had no previous computer experience.

The familiar typewriter-style keyboard makes it easy to enter your instructions. And your programs can be stored on—and retrieved from—audio cassettes, using the built-in cassette interface, so you can swap with other Apple II users.

You can create dazzling color displays using the unique color graphics commands in Apple BASIC. Write simple programs to display beautiful kaleidoscopic designs. Or invent your own games. Games like PONG—using the game paddles supplied. You can even add the dimension of sound through Apple II's built-in speaker.

But Apple II is more than an advanced, infinitely flexible game machine. Use it to teach your children arithmetic, or spelling for instance. Apple II makes learning fun. Apple II can also manage household finances, chart the stock market or index recipes, record collections, even control your home environment.

Right now, we're finalizing a peripheral board that will slide into one of the eight available motherboard slots and enable you to compose music electronically. And there will be other peripherals announced soon to allow your Apple II to talk with another Apple II, or to interface to a printer or teletype.

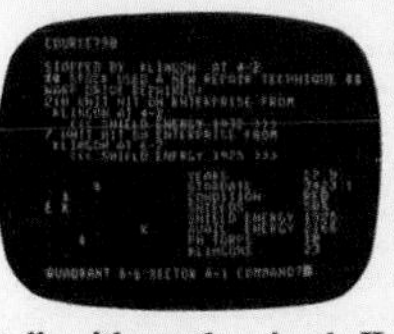

Apple II is designed to grow with you as your skill and experience with computers grows. It is the state of the art in personal computing today, and compatible upgrades and peripherals will keep Apple II in the forefront for years to come.

Write us today for our detailed brochure and order form. Or call us for the name and address of the Apple II dealer nearest you. (408) 996-1010. Apple Computer Inc., 20863 Stevens Creek Boulevard, Bldg. B3-C, Cupertino, California 95014.

Apple II™ is a completely self-contained computer system with BASIC in ROM, color graphics, ASCII keyboard, lightweight, efficient switching power supply and molded case. It is supplied with BASIC in ROM, up to 48K bytes of RAM, and with cassette tape, video and game I/O interfaces built-in. Also included are two game paddles and a demonstration cassette.

SPECIFICATIONS

- **Microprocessor:** 6502 (1 MHz).
- **Video Display:** Memory mapped, 5 modes—all Software-selectable:
  - Text—40 characters/line, 24 lines upper case.
  - Color graphics—40h x 48v, 15 colors
  - High-resolution graphics—280h x 192v; black, white, violet, green (12K RAM minimum required)
  - Both graphics modes can be selected to include 4 lines of text at the bottom of the display area.
  - Completely transparent memory access. All color generation done digitally.
- **Memory:** up to 48K bytes on-board RAM (4K supplied)
  - Uses either 4K or new 16K dynamic memory chips
  - Up to 12K ROM (8K supplied)
- **Software**
  - Fast extended BASIC in ROM with color graphics commands
  - Extensive monitor in ROM
- **I/O**
  - 1500 bps cassette interface
  - 8-slot motherboard
  - Apple game I/O connector
  - ASCII keyboard port
  - Speaker
  - Composite video output

Apple II is also available in board-only form for the do-it-yourself hobbyist. Has all of the features of the Apple II system, but does not include case, keyboard, power supply or game paddles. $598.

PONG is a trademark of Atari Inc.

*Apple II plugs into any standard TV using an inexpensive modulator (not supplied).

apple computer inc.™

Circle 272 on inquiry card.

Plate 14. Cover of the "Domesticated Computers" issue of *Byte*, January 1980. From the collection of Jason Scott.

Plate 15. Cover of *Family Computing*'s first issue, September 1983. Image from the collection of Jason Scott.

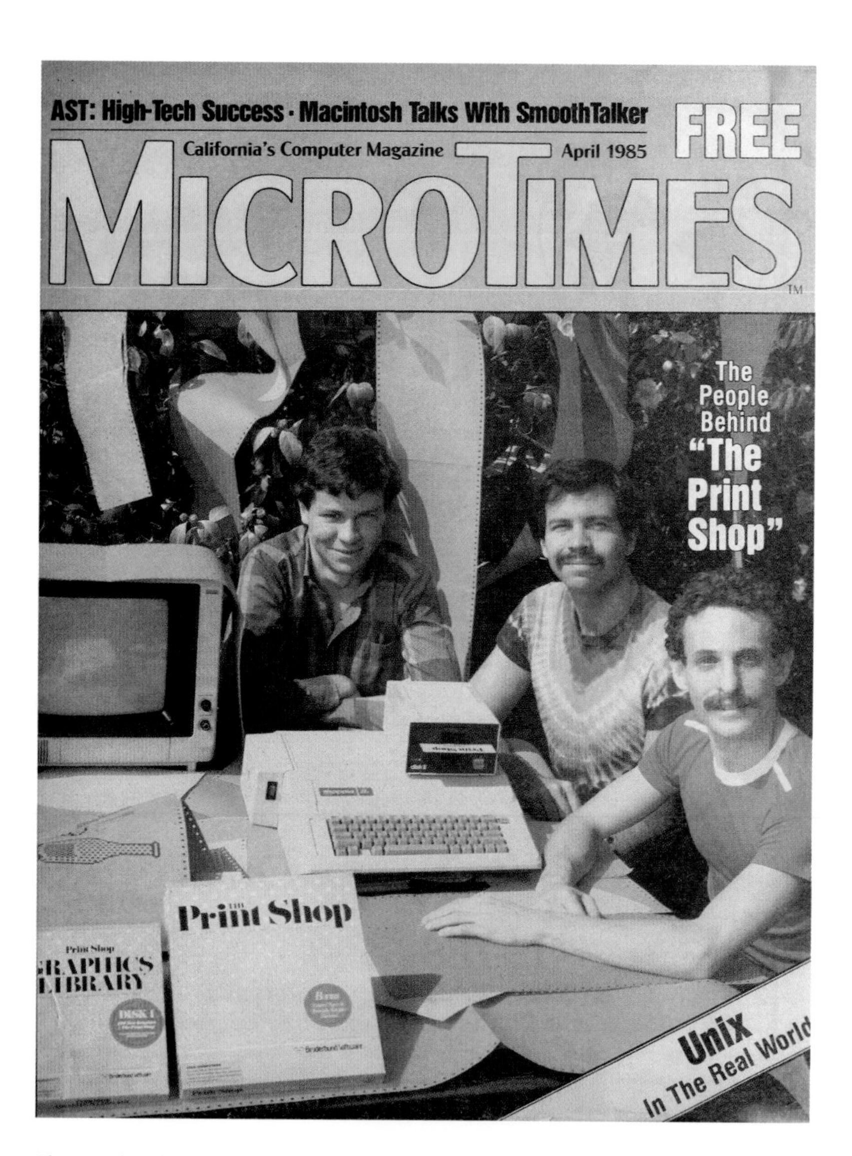

Plate 16. *The Print Shop* and its creators, David Balsam (right) and Martin Kahn (middle), on the cover of *Microtimes*, April 1985. On the left is programmer Corey Kosak. From the collection of Mary Eisenhart. Scan courtesy Jason Scott.

Plate 17. One of Spinnaker Software's first advertisements, showcasing their wide product line of educational software for children. Published in *COMPUTE!*, October 1982, 3–4. Image from the collection of Jason Scott.

Plate 18. Two examples of Spinnaker Software's expanded marketing campaign for its educational products. Ads both published in *Family Computing*, September 1983, 39, and February 1985, 7, respectively. Images from the collection of Jason Scott.

Plate 19. Four Apple II 5.25-inch floppy drives, put out for sale on the consignment floor of the 2021 Vintage Computing Festival East. Photo taken by the author, October 9, 2021.

such programs were not suited to total beginners, or to those without means, insofar as they both required 48K memory—which was only just becoming industry standard at the time and still required a considerable financial outlay.

These two products highlight an important aspect of utility software as a category: the way it supported a robust development culture centered on the Apple II. In a world where microcomputers were competing head-to-head for consumer dollars, a piece of hardware was only as good as the software that was made for it. Some programs, such as *Apple-Doc*, were practical tools for professional programmers. Others, like *Bill Budge's 3-D Graphics System*, encouraged amateurs to imagine themselves as advanced users; as the program's manual promised, "even novice programmers can create impressive animated graphics."[12] Not everyone who picked up these tools would become the world's next software superstar. Nonetheless, these tools did successfully support independent software production for both sale and personal use while also easing the starting curve for amateurs with more productive ambitions.

Unlike more aggressively commercialized areas of the Apple II software market, utilities publishers were almost entirely outgrowths of individual developers (rather than computing marketers, retailers, or from outside the industry), and they often retained a highly user-oriented attitude to software development, marketing, and distribution. Software in this category tended to be well documented and reasonably priced. Aware they were writing for other programmers, developers' reputations were at stake when they released software they wanted respected within a small community of expert users.

Utility software providers also typically took a suspicious outlook on copy protection, distinguishing them from the rest of the commercial Apple II software market. Some utilities publishers, such as Penguin Software and Beagle Bros, did away with copy protection altogether or never implemented it to begin with. Southwestern Data Systems, developer and publisher of *Apple-Doc*, customized a system for users to make a limited number of backup copies. These were not businesses that attempted to predict trends, run pages of color advertising, saturate the market, or chase venture capital. They typically started small and stayed small. A *Softalk* profile of the popular utility manufacturer Beagle Bros focused on the company as a husband-and-wife operation, even three years and several best sellers into its existence (the company's only "semiemployee" was the founder's niece, who handled company correspondence under the name "Minnie Assembler").[13] Thus, even in their industrial practices, these were companies guided by a dedication to the ethos of hobbyist computing, a commercial instantiation of the spirited, hypercurious enthusiast culture that marked spaces like the Homebrew Computer Club and many other early user groups. The developers and publishers who created such software would not be known as software giants, but some would, certainly, make a living.

The utility software category changed little as Apple II ownership expanded into the mid-1980s. While utility software retained a minor amount of popularity within the larger Apple II software ecosystem, utilities rarely accounted for more than two or three best sellers in any given month. In June 1982, *Softalk* parceled out a top ten list dedicated solely to utility software to provide more detailed information on the category (a practice the magazine had already adopted for business software, word processors, and various game subgenres). The list was called "Hobby 10," a nomenclature that identified the list as software for hobbyists—that is, advanced computer users—instead of suggesting a division between hobbyists and professionals, as we might presume today (the list would be renamed "Utility 10" in *Softalk*'s July 1984 issue).

Compared to the proliferation of publishers in hypercompetitive markets like games and business software, utilities publishers remained small in number; given the narrowness of utilities' intended consumers, people did not identify this software category as part of the microcomputing gold rush. Thin competition also meant that a small number of companies retained their significance over months if not years. The mom-and-pop operation Beagle Bros dominated this category from 1982 on, largely due to the sheer volume of quirky utility collections they released, such as *DOS Boss* and *Apple Mechanic*.

Their best sellers landed alongside a mix of releases from Penguin Software, Quality Software, Apple Computer, and the occasional incursion from Phoenix Software, On-Line Systems, and *Locksmith*'s publisher, Omega. The types of software these companies sold also remained relatively stable: programming and graphics utilities, assemblers, programming languages, and various software editors. Even if these products were considered essential or were significantly profitable, they did not garner much curiosity from computing journalism and hobbyist outlets. Utilities were in the business of fixing problems, not creating new modes of computing.

* * *

In the social, economic, and technical world of utilities publishing, *Locksmith* was something of an outlier. This wasn't software just for specialists or programmers. It was a utility aimed at solving a problem for the everyday user—suggesting a connotation for the term *utility* in keeping with notions of public infrastructure, common accessibility, and serving the public good. But what made *Locksmith* useful was what also made it controversial: the ease with which it allowed someone to duplicate programs that software publishers were actively trying to prevent from being duplicated.

Understanding why software duplication was such a thorny issue during the early years of the commercial microcomputer software industry requires understanding the status of software copyright in the 1970s and 1980s, as well as setting aside internet-era notions of piracy. Whether our point of entry to a computational world was desktop computers or touchscreen smart phones, computer users of the twenty-first century have become accustomed to a computing environment where access to software is negotiated through proprietary, often cloud-based platforms—app stores by Tencent or Apple, Valve's PC gaming distribution platform Steam, Google Drive's Docs or Sheets, or Adobe's Creative Cloud. Copying such software is difficult if not impossible, given that the software's code is not physically in our possession.

Personal computers functioned differently in the 1970s and 1980s. Unlike domestic computational technologies such as video game consoles or pocket calculators, the Apple II and many of its competitors were not designed as proprietary or closed systems. Indeed, the entire appeal of a personal computer was that it put computing power directly in the hands of users. It was essential that users be able not only to program on their machine but also to save and distribute the work they did. These dynamics are, in part, what distinguished microcomputers from game consoles, on which the user could only ever consume software rather than create it themselves. Such dynamics were also essential to building a software market that would compel the purchase of this inordinately expensive hardware. Consequently, flexible storage formats were desirable. Cassettes and floppy disks were relatively inexpensive, relatively durable, and gave users control of data. But that control cut both ways, since it gave users the ability to copy not just their own work but anyone else's too.

The question of software duplication went straight to the question of software's status as a creative work, which also defined its legal status. Since the 1960s, software had been acknowledged by the US government as protectable under copyright as literary works. However, this protection was not extensively used within the mainframe and minicomputer industries, which were generally more interested in protecting novel programming ideas than in preventing outright duplication (as opposed to the book, music, and film industries).[14] Software piracy was also less common in the 1960s and 1970s because of the close contact between businesses and their hardware manufacturers, as well as the customization required to implement software in any given installation.

Circumstances emerged differently in the early microcomputing industry. Many hobbyists embraced software as a "communally produced and jointly owned good," developing, copying, and sharing useful programs in periodicals or hobbyist peer groups.[15] On the other hand, for-profit software providers (including hardware manufacturers, software publishers, and self-publishing developers) typically exercised copyright claims to protect their economic interests—in particular, to prevent duplication.

In an effort to dissuade those who might duplicate software to give away or sell to others, software publishers placed copyright notices in program documentation. Such notices were typically idiosyncratic, varied widely, and, at least in the mid- to late 1970s, were rarely backed up with a formal filing at the US Copyright Office.[16] For example, the copyright notice that accompanied Peter Jennings's self-published 1976 KIM-1 game *Microchess* specified that the program and its documentation were "provided for the personal use and enjoyment of the purchaser."[17] While Jennings's game provided a source code listing that invited players to "expand or modify" the game "to suit the requirements of [their] particular system configuration, or to experiment with [their] own ideas for improvement of the playing strategy," users were nonetheless expressly prohibited from "reproduction by any means."[18] In other words, you could tinker with the code for your personal amusement, experimentation, or education, but you could not reproduce the original code or any alterations derived from it. However, once Jennings teamed up with Dan Fylstra at Personal Software (*VisiCalc*'s future publisher) in 1978 to release *Microchess* 2.0 for the Apple II, these prohibitions became more detailed in both scope and consequences. The following notice was printed on the first page of the game's manual:

> Microchess 2.0, a computer program for the Apple II, is copyright © 1978 by Micro-Ware Ltd., and is published exclusively by Personal Software Inc. Copyright infringement (the making of copies on disk or tape, in printed form, via communication lines, or by any other means) is a crime punishable by fines and imprisonment, regardless of the number of copies made and whether or not a profit motive is involved. In addition, infringers may be sued for civil damages. Micro-Ware Ltd. and Personal Software Inc. will seek maximum criminal penalties and damages against persons violating the copyright laws.[19]

In firming up the copyright for *Microchess*, Personal Software also removed users' ability to edit or modify the game legally—a resounding death knell for the hobbyist impulses of exploration and discovery that undergirded Jennings's original creation. While it is unclear whether such claims would have been enforceable under copyright law as it was understood in 1978, the intended effect was clear: to establish financial and legal consequences for any act of duplication.

Software duplication concerned the microcomputer software industry for the same reason that library photocopiers worried book publishers in the 1960s and VCRs raised fears within the film industry in the 1980s: ready access to reproduction technologies threatened the economic hold publishers and other stakeholders had on the market. The arrival of microcomputers as a personal technology, located in domestic spaces or private offices, enabled acts of software reproduction that could not be traced, tracked, or otherwise surveilled.[20] Furthermore, the "Hacker Ethic" prevalent among early microcomputer hobbyists, embodied in notions like "all information should be free" and "mistrust authority," meant some users considered piracy a moral obligation for spreading the utility of personal computing.[21] As discussed in chapter 1, the rampant circulation of Bill Gates and Paul Allen's 1975 *Altair BASIC*, duplicated on paper tape and given out for free at user group meetings, was an early indication that some hobbyist subcultures did not share the ideological interests of entrepreneurs who wanted to replicate capitalist modus operandi within the emergent US microcomputing industry.

* * *

As the unauthorized duplication of *Altair BASIC* demonstrated, copyright-based prohibitions against software copying were relatively unenforceable, especially if such practices were hyperlocal and noncommercial. This was especially true as commercial software transitioned from paper tape onto more easily replicable formats such as cassettes and, soon after, floppy disks. The prevalence of at-home cassette decks (and, by 1979, mass-manufactured floppy disk drives) made copying data a relatively trivial procedure. Indeed, the capacity to copy data from one floppy disk to another was considered such an essential and obvious component of data management that copy routines were included as standard commands in all microcomputer disk operating systems (DOS), including Apple's Disk II.

In response, commercial copyright holders began quickly exploring technical mechanisms to prevent duplication from one cassette or floppy disk to another or at least to make it more challenging. The earliest known uses of techniques to disrupt a straightforward copying process date to 1978. Cassette-based Apple II games such as *Microchess 2.0*, *Sargon II*, and Softape's *Module 6* have all been found to include various programmatic tricks that encoded data to the tape and assigned tape data to system memory in ways that prohibited straightforward, tape-to-tape copying.[22]

As the industry transitioned from cassettes to disks, however, software publishers and developers began devising far more sophisticated methods for manipulating the way data was stored and read. Copy protection flourished in the early 1980s, becoming so widespread and so ingenious in implementation that it could be said to constitute a kind of vernacular industrial programming practice unique to floppy disks.[23] As discussed in previous chapters, the Apple II's floppy disk drive, the Disk II, was released late in 1978 and took off following the release of *VisiCalc* in 1979. Floppy disks were faster at locating and loading data than cassette tapes because while tapes required data to be accessed sequentially, floppies allowed data to be accessed at random anywhere on the disk. This random-access capability was facilitated by an organizational schema that divided the disk into 35 tracks and 16 (originally 13) sectors.[24] Tracks are concentric rings moving outward from the center hole of the disk; sectors are divisions that cut radially across the tracks, often depicted as the slices of a pie (although such a depiction was not entirely accurate) (see chap. 3, p. 85).[25] Many other "rules" guided how an Apple II floppy disk was organized: each sector was designed to hold 256 bytes; there were intentional empty bytes between sectors, and a special sector data marking, "D5 AA 96," was used to indicate to the Disk II drive where new sectors started.[26]

Importantly, the fact that an Apple II floppy disk worked this way—that it was divided into 35 tracks and 16 sectors, with 256 bytes per sector, D5 AA 96, and all the rest—was not dictated by the physical features of the floppy disk itself but rather by the Disk II's DOS, as well as the specific hardware configuration of the Disk II's peripheral card. DOS was *software*, which came on a floppy disk called the System Master that was packaged with the Disk II.[27] Its job was to manage the flow of data and execution of operations between the drive and the computer. DOS was where all the BASIC commands for controlling the disk drive were kept; loading these commands into memory from the DOS System Master was known as "booting the disk" and was thus the first essential step in using the Disk II. This meant the DOS System Master disk was required for activities as innocuous as formatting blank disks (the INIT command), viewing the contents of the disk (CATALOG), or copying the contents of one disk to another (COPY or COPYA, depending on the version of BASIC in use).[28]

Copy protection worked by manipulating the gap between the data architecture rules that were part of the DOS System Master copy routines, COPY and COPYA, and the fact that a skillful programmer could save data to a disk using any data architecture they desired, so long as the instructions for interpreting that architecture were also saved to the disk. When an Apple II user directed DOS to copy a disk, the copy routines assumed a standard data architecture: 35 tracks and 16 sectors, 256 bytes per sector, and so on. In essence, the DOS copy routines used the standard architecture as guidance for how to move data from the disk being copied into the Apple II system memory and onto a new disk. Yet nothing required a developer or publisher to organize a disk's data architecture according to the standard system Apple had implemented. Through a variety of software techniques, developers and publishers could defy the "rules" of how disks were supposed to work and instill their own rules in the disk's bootstrap code. Such manipulations are numerous, including everything from placing tracks between other tracks to changing the special sector data marking to moving the disk directory out of track 17 (where it was usually held to minimize movement of the read/write head) to laying out the data in a spiral pattern across the surface of the disk rather than in tracks.[29] Such bespoke data layouts didn't create problems when loading or running a disk, but they did make it impossible to *duplicate* the disk using the COPY/COPYA command, because DOS's copying commands presumed the standard data layout. In other words, nothing in DOS's copy commands could assess if a disk was saved in a nonstandard format. If data were not where they were supposed to be, the copy operation would fail, resulting in either an error message or a disk that appeared to have been successfully duplicated but would not actually work.

So while one's personal disks and self-written programs would have been easy to copy using DOS, publishers of commercial software were able to use such programmatic techniques at scale to effectively prevent consumers from making duplications of store-bought software. Copy protection could literally follow consumers out of the store and into their homes.

* * *

The circumstances under which "unauthorized" software duplication became a concern in the early microcomputing era stemmed from multiple sources: a failure of ideological consensus about the status of software as intellectual property, the economic anxieties of an emerging class of software entrepreneurs, and the very material challenges copy protection created for everyday users. The goal of most copy protection efforts was not to prevent all piracy, which was technologically impossible; as one early computer user put it, "For any copy-protection scheme devised, some bright person somewhere will invent a way to bypass or break the protection."[30] Instead, the aim was to make copying commercial software challenging enough to dissuade casual duplication, the sort of escalating consumer bogeyman scenario outlined in industry-sympathetic magazines like *Softalk*:

> Starting by making copies for enthusiastic friends, some personal computer users move on to cranking out tens to hundreds of copies that they nonchalantly pass on to their friends' friends and mere acquaintances. Where scruples draw a line and a halt varies; but few who have gone this far can resist the opportunity to profit from their work, and they begin to offer "their" product for sale.[31]

In the same feature, *Softalk* journalists extrapolated that if the average Apple owner possessed $100 worth of pirated software and Apple was gaining ten thousand new customers a month, then $1 million was being "siphoned out of the industry monthly."[32] This was the general party line most publishers, as well as mainstream journalists, took with regard to copy protection, insisting that without ways of preventing unauthorized duplication, "illegitimate copies of programs threaten the fabric of personal computing."[33]

Unlike film studios or recording companies, which were American media behemoths, personal computing entrepreneurs appealed to the underdog sentimentality of early users, insisting that the only way to ensure a healthy industry, and the growth of personal computing, was to make sure programmers and publishers were fairly reimbursed so they could keep making innovative and necessary products. Following this logic to its inevitable conclusion, entrepreneurs like the On-Line Systems cofounder Ken Williams insisted that "the most adverse effect of piracy is not so much the higher price of software as it is the lower quality of new programs," suggesting that piracy would lead talented programmers to exit the industry, leaving only "weekend programmers" making low-quality products.[34] In such an economic framework, copy protection techniques were put forth as a form of loss prevention that was essential not just to protecting individual products, but to saving an entire industry.

Rationales for engaging in unauthorized duplication were complex. Users, such as their opinions can be documented, shared concerns about the limitations copy protection placed on both the practical usability of their software and the ideological, Hacker Ethic–style concerns related to software ownership and free access of information. Anecdotal and archival evidence indicates that there were explicit piracy rings large and small, often running out of local users groups or even software retailers.[35] Some users were not keyed in to the software journalism scene well enough to realize that disk duplication broke the law; others insisted no harm was being done as long as they didn't charge for the copied software or that they could do whatever they pleased with their property. And some pirates ascribed to the notion that software prices were unreasonably high, especially when it was clear that some publishers were already raking in millions. And for school districts, the sheer expense of buying so many copies of an individual piece of software could be prohibitive, leading one teacher to wonder if a clause allowing users to make backup copies for personal use would extend to "a school system purchasing a package and then copying enough for all the schools in the system."[36]

The knock-on effects of copy protection produced a far more material set of grievances among consumers. Preventing disk duplication also prevented users from making their own backup or archival copies of their software.[37] Creating backup disks was a routine and often necessary practice for floppy disk users, ensuring they had a replacement in case any crucial data on the original disk became damaged. The fragility of floppy disks was significant enough that the Apple II DOS manual dedicated an entire section of its fourth chapter to "protecting yourself against disaster." As the manual states:

> Floppy disks are sturdy and reliable. . . . But it's still possible to lose or destroy all information on a diskette. A diskette may get scratched or damaged by heat; it may get lost, or a dog may chew it; someone may decide to use it as a frisbee at the beach; if a diskette isn't write-protected, it may accidentally get written over. And a diskette will eventually wear out—a lifetime of 40 working hours is about average.
>
> ** Moral **
>
> Keep more than one copy of a program around if you don't want to lose it. In computerese, "back up" any valuable program.[38]

Such claims were hardly hyperbolic; stories abound of the unusual circumstances floppy disks found themselves in, forming a kind of micro-genre among early computer users. One *Softalk* reader documented blow-drying a floppy disk after his children spilled water on it, and another recounted having his disk get covered in ice cream after storing it in a kitchen cupboard during a party.[39] One truly remarkable letter detailed a Rube Goldberg-esque set of events that resulted in their floppy disks getting covered in suntan lotion *and* falling out of the back of a moving truck (with a few Q-tips and some rubbing alcohol, everything booted just fine).[40] Leaving disks on refrigerators, microwaves, or even too close to the computer itself could disrupt their magnetic field, while problems with disk drive alignment or maintenance or even the failure to use a surge protector could result in data overwriting. Programs themselves could have data access errors that caused problems using software features or cause the disk to infinitely spin in its drive, increasing the likelihood of damage.[41]

Consumers who suddenly found themselves with a botched commercial disk and no backup had limited options. Publishers were generally amenable to replacing damaged or destroyed disks for customers who mailed in the unusable versions of their products, but policies varied. *VisiCalc* was fairly notorious in the early 1980s for initially charging $30 for a backup disk, along with requiring the warranty card.[42] Yet user errors were common, comprising 13 percent of a random sample of malfunctioning disks returned to the company Sir-Tech Software, which included forms of physical damage such as "peanut butter and jam particles, pencil marks, paperclip impressions, pinholes, and severely creased disks."[43] Similarly, customer service personnel at Brøderbund Software were so amused by a returned disk mauled by a dog that they photocopied the disk and circulated the image in their internal newsletter.[44] Yet the process of getting replacements could take weeks or months, in addition to whatever costs the publisher might charge. While such issues might have been merely an inconvenience for users unable to play their favorite game, extended inaccessibility to programs like word

A photocopy of a 5.25-inch *The Print Shop* floppy disk that had been chewed up by a user's dog. The disk itself was mailed to Brøderbund as evidence of damage, in order to receive a replacement disk. Brøderbund employees published a photocopy of the disk in the October 1984 issue of the company newsletter. Image courtesy The Strong, Rochester, New York.

processors, spreadsheets, or database managers could wreak havoc for businesses.

* * *

This tangle of conflicts between publisher and consumer desires was exacerbated by the enactment of the Software Copyright Act of 1980, passed by Congress that December as an amendment to the Copyright Act of 1976. As the computing historian Gerardo Con Díaz has noted, "The act single-handedly established computer programs as a new class of copyright eligible works" rather than treating software as simply an appendage of literary works.[45] Importantly, and pursuant to software's function as "a set of statements or instructions to be used directly or indirectly in a computer," this amendment to the Copyright Act also entitled software owners to make backup or archival copies of their disks—a right consumers had not explicitly had previously.[46] In other words, the 1980 Copyright Act was the condition of possibility for *Locksmith* and similar competitors like *V-Copy*, *Copy II Plus*, and *Back-It-Up* to safely circulate as commercial products.[47] For *Locksmith*'s president, Dave Alpert, the Copyright Act "legitimiz[ed] bit-copy programs. We had been in the process of coming out with our program and when we saw that . . . we knew we were there."[48] Similarly, *InfoWorld* cited the Software Copyright Act of 1980 as the catalyzing event for the small but threatening disk duplication market, noting, "Since [December 1980], several programs that can copy 'uncopiable' programs, encoded by various copy-protection schemes currently used by major producers of software for the Apple, have appeared on the market."[49]

Locksmith's earliest advertisements, which ran in January and February 1981, thus rode the wake of the Copyright Act, promising an Apple disk copy utility that could copy the uncopyable.[50] *Locksmith* accomplished this by producing "BIT by BIT cop[ies] of your disk"—in other words, copying data one bit at a time, regardless of how it was organized on the floppy, rather than assuming a standard data configuration. This technique allowed *Locksmith* to get around copy protection by producing a bit-level duplicate of the data on disk (leading this general category of utilities to be referred to as "bit copiers" or "nibble copiers"). Importantly, *Locksmith* did not present itself as a utility to facilitate outright piracy. Rather, *Locksmith* claimed its functionality rested in the peace of mind it offered users, who would "no longer . . . have to worry about spills, staples, or magnetic fields that destroy [their] valuable diskettes," so long as they duplicated their disks with the program.[51] *Locksmith* based its sales pitch (and its $74.95 price point, or roughly $215 in the early 2020s) on the premise that the program would pay for itself in the time and money it saved users from having to wait for replacements or pay extra fees in order to obtain backup copies. *Locksmith*'s agenda was to offer computer users a tool to regain forms of control over their software that publishers desperately wanted to inhibit.

Relatively little is known about the provenance of *Locksmith*, or its publisher, Omega MicroWare (advertised variously as Omega Software Systems, then Omega Software Products, during its first several months of existence). According to a history provided in the program's 1983 manual, *Locksmith* had initially been programmed as an educational exercise called NIBY, not originally intended for sale; the nibble copier garnered interest from the programmer's local Apple users group, which pushed for the project to be marketed.[52] Yet Omega never publicized the name of the software's programmer. In a 1981 interview, Omega's president, Dave Alpert, stated, "I am not allowed to talk about the author." But the 1983 *Locksmith* manual claims the utility was developed by "an Apple programmer with 18 years of computer experience, including systems programming on large IBM mainframes at several large corporations."[53]

Locksmith's relative absence in the historical record is, of course, a by-product of the industry's collective decision to blacklist the program. Computing enthusiast magazines did not run features on Omega or its infamous program, likely due to concerns about advertiser pushback, leaving a conspicuous gap in a magazine world that otherwise typically covered every minute shift and change and advance in the microcomputer software and hardware markets. Yet in the negative space of *Locksmith*'s shadow, that same magazine world paradoxically left a great deal of commentary about its decision to *not* run *Locksmith*'s ads. Robert Tripp's editorial apology in the pages of *MICRO* was only the beginning; the entire microcomputing industry journalistic establishment became part of the fray, as editors at various outlets checked in with one another and debated with their own editorial staff about what position they should hold on advertising bit copiers. Documenting some of this interplay in *COMPUTE!*'s March 1981 editorial on copy protection, editor and publisher, Robert Lock, noted an exchange with *MICRO*:

> Several other magazines in the industry have recently been running advertisements for a program that copies (duplicates) protected software for a particular machine. One of these magazines (Micro) checked with us to see our feelings on such advertising. We indicated that we wouldn't run such ads, given that the software in question could be used to produce duplicate copies of "protected" and proprietary software. We understand that Micro has since decided to do the same. We applaud this move toward protecting the rights of the software industry, and encourage additional comment.[54]

COMPUTE! was toeing a careful line. Despite insisting on the need to protect the "rights of the software industry," Lock also wanted to insist that the statement "isn't intended to be an inclusive comment on the rights of software buyers," acknowledging that copy-protected disks did cause problems and that vendors needed to have a more "customer oriented, fully responsive plan for allowing licensed owners of software to quickly, conveniently and most of all, economically, obtain a backup in case of failure of a diskette."[55]

InfoWorld followed suit a month later, titling their April 13, 1981, editorial, "No More!" Their stated position was to not only refuse to accept advertising for bit copiers, but to "not review products that only make copies of protected programs . . . [or] publish articles or information that will allow the user to unlock the copy protection scheme used on software."[56] Like *COMPUTE!*, *InfoWorld* acknowledged that the software industry itself was partly responsible for driving demand for copying programs but nonetheless set its policies clearly to favor the industry rather than leave the matter up to consumers.[57] In the commentary of various editors and microcomputer journalists, it's possible to confirm that almost every magazine in the business had either explicit or implicit prohibitions against advertising bit copiers, including not just *MICRO*, *COMPUTE!*, and *InfoWorld* but also *Creative Computing*, *Call-A.P.P.L.E.*, and *Softalk*.[58]

However, the deepest historical insight into this morass of disagreement comes from *Hardcore Computing*, a magazine launched, in part, as a direct response to the censorship of *Locksmith* by more mainstream venues.[59] Publishing its first issue in summer 1981, *Hardcore Computing* was dedicated to "Apple-users . . . who are tired of being kept ignorant by the other magazines and told that it is for their own good."[60] *Hardcore* ran a smattering of opinions, interviews, reviews, tricks, and hacks, alongside extensive documentation on how to crack disks with increasingly sophisticated software protection. Revealing inside details on how to break copy protection was necessary, according to *Hardcore*'s publisher, Chuck R. Haight, because there was a "raging, silent battle between Apple-users and the magazines."[61] As Haight wrote in his opening message to *Hardcore*'s inaugural subscribers:

> This problem became more apparent when "Locksmith", a bit-copier that would make duplicates of many copy-protected diskettes, was censored (the magazines refused to publish the ad, thereby denying their readers the knowledge of the existence of such information). That was censorship! And the battle was on. . . . Any magazine that took a stance against the Apple-users could not be a magazine for me. . . . Because there was no magazine out there for me, I decided to fill that void myself.[62]

Hardcore's editor, Bev R. Haight, similarly referenced the *Locksmith* incident as an impetus for *Hardcore*'s creation, stating, "If you don't know about Locksmith, then their editorial campaign of ignorance has succeeded."[63] The magazine also provides the only known review of *Locksmith*, evaluating it against its most prominent competitors, *Back-It-Up* and *Copy II Plus*. Overall, the program was regarded as "an elegant solution" and "user-friendly," with "clear, informative, and professional" documentation.[64] Of all the bit copiers evaluated, *Locksmith* succeeded in copying the most pieces of software in the review test, twenty-three of twenty-five (the next best being twenty of twenty-five).

Beyond the magazine's editorial posture, however, *Hardcore* took its defense of *Locksmith* even further, featuring an extensive two-page interview with Dave Alpert, described in the magazine as simply "head of Omega Software, Inc."[65] In the roving conversation, Alpert is given extensive license to tell his side of the story, including making claims about the actions and private conversations of other editors, among them, Robert Tripp of *MICRO* and Val Golding, editor of the Apple user group magazines *Call-A.P.P.L.E.* and *Apple Orchard*.[66] In Alpert's account, he accuses Tripp of not just checking in on the decisions of other editors, but of "[telling] them that if they ran our ad, they would lose business. He went on a campaign to spread this information."[67] In Alpert's telling, "users around the country" had criticized Tripp's decision: "I had people call me and say that they sent letters to Tripp telling him that they wanted to boycott Micro or that they wanted to boycott the people who advertise in Micro. . . . I received hundreds of letters from *Locksmith* owners telling me how grateful they are to have the ability to make legitimate back-ups."[68] But Alpert's most virulent accusations were leveled at Golding. Alpert claimed that Golding, a prominent member of the Apple Puget Sound user group, was so dedicated to ensuring *Locksmith* ads didn't run that he arranged for his user group's members to pull all their articles from *Apple Orchard* if it published the *Locksmith* ad.[69] In Alpert's estimation, the behavior of editors like Tripp and Golding was not just unscrupulous; it was a violation of the law, an "attack upon our company and upon our ability to advertise a legal product."[70] Reporting from *InfoWorld* in May confirms that Alpert had successfully used legal sanctions to make at least one magazine, *Nibble*, run *Locksmith*'s advertising.[71] Eager to fan the flames of this industry disagreement, *Hardcore Computing* published an entire one-page rebuttal from Golding, who claimed Alpert's story was inaccurate and based on hearsay.[72]

The question of truth here is lost to time—and perhaps isn't the point. More compelling is the way these industrial circumstances shaped knowledge of this product in both the present of the 1980s and the future in which it would be written about. By establishing a party line against the circulation of *Locksmith*, early 1980s publishers created a tremendous chain of evidence about the opposition to *Lock-*

smith while simultaneously preventing more extensive information about the company itself from landing in the historical record.

Blacklisted from magazines, the program circulated largely through word of mouth, with some retail coverage. As Alpert claimed, by summer 1981 he had already sold thousands of copies, but believed "I could've sold 3 or 4 times as much had I been able to advertise."[73] Alpert likely wasn't wrong; word of mouth was effective enough to push the software into retail outlets, where it subsequently rose to the top half of *Softalk*'s Hobby best-seller list by the end of 1981 (despite the prohibition against *Locksmith* advertisements, the magazine wasn't willing to alter its sales reporting). *Locksmith* would flicker off and on that list for the next two years, usually cropping up in the first several months of 1981 and 1982—reflecting seasonal holiday buying patterns.[74]

Yet the product's popularity among Apple II devotees belies its scattered appearances in the Hobby Top 10. In 1982, *Softalk* launched an annual reader's choice feature, based on subscriber rankings of their favorite Apple II programs released the previous year, as well as their favorite programs of all time. *Locksmith* ranked third among the most popular utilities released in 1981—outbid only by utilities released by Apple Computer itself, including its DOS 3.3 update. In other words, *Locksmith* beat out every other third-party utility program on the market. Overall, the bit copier ranked twenty-first of the thirty most popular programs of 1981.[75] For the 1982 rankings, *Locksmith* shared a three-way tie for first place in the Hobby category with *Global Program Line Editor* and *Graphics Magician*, placing twelfth in the all-time Apple II software category (one of only four utilities to make that list). Most astoundingly, however, *Locksmith* came in third place for the best program released in 1982, falling between two hit games from that year, *Wizardry* and *Cannonball Blitz*.[76] Yet this strong showing in *Softalk*'s annual popularity contest is inconsistent with *Locksmith*'s uneven standing on the best-seller list—either suggesting that a large number of users were registering popularity for a product they didn't buy or, more likely, signaling that the majority of

Locksmith revenue was being taken in by Omega MicroWare through direct sales, thus leaving its respective economic strength indecipherable within the overall Apple II software market.

While the top magazines of the US microcomputing enthusiast world might have been willing to boycott *Locksmith* ads due to ambient pressure from advertisers, they wouldn't blacklist the company as a whole. From fall 1981 on, Omega MicroWare was able to run numerous advertisements for less controversial products, including other kinds of memory utilities, tax preparation and financial investment software packages, RAM expansion boards, and even a game called *Night Falls* (designed by none other than *Locksmith*'s stalwart defender at *Hardcore Computing*, Bev R. Haight).[77] In fall 1981, Omega MicroWare was able to get away with running an "advertisement" for *Locksmith* that essentially served as an extended op-ed on the question of copy protection in several major venues, including *Creative Computing* and *BYTE*.[78] The format is curious. Rather than take the form of an open letter or a letter to the editor—a more typical genre within the technical and textual communities that constituted early microcomputer users—the announcement is sized like a half-page advertisement and, in some instances, crosslisted in the advertisers index (an almost baiting gesture toward software vendors who had opposed *Locksmith* advertising). Offering "our side of the story," the ad lays out considerable background on Omega's decision to publish *Locksmith*, citing the Software Copyright Act of 1980 and reminding readers of their legal right to create backup disks: "It is now the law of the United States that the existence of a copyright notice on a computer program does NOT make it illegal for the legitimate owner of that program to copy it for archival purposes." Backing off from the kinds of opinions Alpert expressed in *Hardcore*, the enemy established in the advertisement is not magazine editors but publishers who failed in their duty to "educate the public as to their rights and responsibilities" and instead chose to "pressure magazine publishers into refusing our advertising, and to invent new copy protection schemes."

It is unknown what rationales magazine editors used to publish this advertisement, whether legal force on the part of Omega MicroWare was involved or whether this advertisement provoked further pushback from software vendors. What is clear, however, is that the issue of a consumer's rights with regard to copy protection was not going down quietly. After all, many editors had mentioned in their public statements that the industry itself was partly responsible for creating the very conditions that primed users to seek remedy from copy protection schemes. In May 1981, *BYTE* devoted its issue cover and features to the topic of software piracy, opening with an editorial that asked, "How does the manufacturer serve the customer's legitimate need to make backup copies, while protecting his expensive software investment?"[79] The answers *BYTE*'s editor proposed are not nearly so telling as the frequent need to ask the question at all.

Locksmith, then, became a catalyst for a much larger set of industrial outcomes. Some interventions were small stakes, opportunistically leveraging anxieties about nibble copiers to spawn further games of software cat-and-mouse. As one example, in the wake of *Locksmith*, a little breed of nibble copier protection programs emerged, including Double-Gold Software's *Lock-It-Up* and Zivv Company's *Disk Protection Program*, promising to protect against programs like *Locksmith*, *Back-It-Up*, and *Copy II Plus*. Yet some industry actors responded at a more structural level. In 1982, a handful of publishers announced they were doing away with copyright protection altogether, posting their announcements via letters to the editor published in venues like *BYTE* and *Softalk*. Mike Pelczarski, president of Penguin Software, a prominent maker of graphics utilities, wrote that while he too was afraid of the damage casual piracy might impose on the sustainability of his business, "our desire to make a good product won (but not by much) over our fear."[80] Operating from an ethic that put his consumers first, Pelczarski hoped that "added convenience will result in more sales, not less, and that the software market has matured to the point where people realize that the result of illegal copying is less convenience for everyone."[81] In a simultaneous move, E. J. Neiburger, president of the health care systems software producer Andent Inc., insisted that in his company's experience, "unlocked software is good for business—

our business, your business, and the customer's business."[82] While controlling large-scale piracy would always prove a challenge, Neiburger felt casual piracy could be circumvented through low costs and the benefits that come with registering one's program for updates and future documentation. Though such stances among publishers were more the exception than the rule, they do show that even publishers were not unified in their attitudes to copy protection or unsympathetic to the experiences of their consumers. These changing attitudes are another penumbra of *Locksmith*'s long shadow—just one more example of an industry seeking equilibrium in the tug-of-war between consumers' desires and publishers' economic motives.

* * *

No one would win the war over disk duplication, but consumers almost certainly lost. The positions of Andent Inc. and Penguin Software would not be repeated by the publishing community at large; there would be no great copy protection stand-down. Publishers would not risk their bottom line to make their software more usable. Certainly, disks would still become corrupted, get accidentally used as frisbees and coasters, stored atop computers or refrigerators, creased between books or filing cabinets or desk drawers. Disk drives would still spin out, overclock, eat their own software. Yet lockout mechanisms would only grow more sophisticated, and publishers would devise endless programmatic curiosities to challenge those who sought to replicate software outside the industry's oversight. Like all the commercial media industries that had come before, from books to music to film, the personal computing software industry would align in its interest that capital must be captured. In the standoff between publishers and consumers, unauthorized duplication was merely a tactic, a disorganized and uncoordinated effort to serve the software industry a thousand paper cuts from below.

What the software industry would devise, however, was a strategy, top-down. In April 1984, a confederacy of publishers, developers, and other industry actors pooled their financial support to launch the Software Publishers Association (SPA), a trade organization headed by the Washington, DC, attorney Kenneth Wasch, to protect and expand the internal interests of the personal computing software sector. By spring 1985, membership totaled over 120 firms, including Sierra On-Line, Brøderbund, Activision, and Spinnaker, as well as the software divisions of traditional media publishing companies such as Bantam, Scholastic, Random House, Reader's Digest, John Wiley & Sons, and Prentice-Hall.[83] Like any trade association, the SPA had aspirations to build its lobbying power, collect industry data, host annual dinners, and deck its membership with awards. But beyond these activities, the SPA's greatest task was to stanch piracy.

While the 1980 revisions to the Copyright Act afforded consumers the right to archival backup copies, they also granted representative bodies like the SPA the authority to pursue litigation against any entity suspected of piracy, on behalf of its membership. No longer would software pirates, professional or casual, be chased after by individual publishers playing a relentless game of whack-a-mole. Instead, the SPA built an entire internal division dedicated to chasing piracy leads and set up a $50,000 copyright protection fund dedicated to the implementation of "a realistic, action oriented program to frustrate illegal copying of microcomputer software."[84] Adversaries included the individuals who copied and resold software for a profit; bulletin board systems that distributed or facilitated the distribution of computer software; user groups, businesses, schools, or universities that used unlicensed copies of software; software rental companies; and firms like Omega MicroWare that "market[ed] software copying devices and programs."[85]

Endowed with the financial support of the software publishing industry, the SPA's strategy was surprisingly bespoke. Rather than drag pirates into court or simply refer them to the FBI, SPA leadership preferred to shake down pirates internally first. The SPA became well known for its threatening letters but was also not above hiring private investigators or even conducting its own "raids." In one instance, Wasch himself posed as a customer at a computer retailer in the New York area that was known for selling copyrighted software off the books, caught the employees in the act, and then worked out a deal with the company owners to not pursue damages in exchange for a $200 donation to the SPA's copyright protection fund.[86]

Such activities were fully supported by the SPA's membership, insofar as publishers whose work was being illegally duplicated signed releases to the SPA stating they would not pursue separate damages. In another instance, SPA followed the lead on a solo pirate working out of Irvine, California, which turned out to be a thirteen-year-old boy who may or may not have been selling illegally duplicated software. Restitution was made in the form of a handwritten letter: "I'm truly sorry about this. . . . I had no idea what 'copyright' meant, or what it is for, or the laws about it. My dad never explained this to me. . . . I realize now what a big mistake I almost made, and I apologize for what I did. I won't ever do this again."[87] What was valuable to the membership was not restitution for lost revenue but the chilling effect such behavior had on piracy. Somewhere, a child learned that the entire force of the federal government could walk through his door if he did bad things with a computer. Once more, the computing industry's capacity for revenue capture relied not on the allegedly radical capacities of the technology or the freethinking impulses of its user base but on the eagerness of its earliest stakeholders to align themselves with government, surveillance, and laws they were well positioned to disproportionately take advantage of.

* * *

If the world of microcomputing might once have been perceived as one in which users and developers were aligned in their interests, it took less than a decade for that to clearly no longer be the case. A pirate was a pirate, not a user or customer. By emphasizing unauthorized disk duplication as an indisputably illegal act, no matter the rationale, the software industry effectively cordoned off a part of the user base as bad-faith criminals unworthy of sympathy. And yet no publishing company or developer ever claimed it went out of business because of piracy. Omega MicroWare, on the other hand, wouldn't survive the mid-1980s. While *Locksmith* did well enough, Omega would peter out over the coming years despite its other products. *Locksmith* transitioned to the publisher Alpha Logic Business Systems by 1985 and stayed on the shelves for a few more years before disappearing into obscurity, along with its new publisher.

Although *Locksmith* remains a ghost in the machine of computing history, its archival residue, thin as it is, provides a trace record of the myriad ways users determined for themselves what the fair and ethical use of software was—or whether they cared. While it may be tempting to think of Apple Computer as merely an accidental bystander in the debates surrounding copy protection, bit copiers, and consumer affairs within the microcomputing industry, it's hardly a coincidence that it was a piece of Apple II software that ignited this conflagration. What made the system ideal for such a broad range of third-party software development was also what made that software library so exploitable for unauthorized duplication, and so attractive to publishers looking to lock in their profits.

Thus, from an industry perspective, *Locksmith* serves as an example of how utilities got to the heart of how people used their computers. Negotiating the boundary between user and machine, utilities were software applications that helped people make the most of the computer itself, whether streamlining code, fixing memory errors, or backing up the software they needed to run their businesses. And yet as a general rule utilities have largely been fated to hum in the background of computer history. *Locksmith*, however, takes us to a moment when this unassuming category of software seemed to carry the stakes for the entire industry. In *Locksmith* we find a history of computing that is precisely about *how* people could use their computers, and a surprisingly human one at that, full of fried disks and overspun drives, heated phone calls between magazine editors, accusations dashed out for the entire industry to see, and people worried about their livelihoods, on all sides. Whatever we may think of these editorial stances, publisher positions, and consumer pleas, it bears remembering that what got everyone talking about *Locksmith* wasn't the software itself—it was the way the program threatened the economic premises on which the entire software industry was building itself.

6

Home

The Print Shop

"Clear the kitchen table." Such is the opening copy for Apple's vibrant Apple II ad campaign, appearing in *BYTE* as early as June 1977 (see pl. 13).[1] The full-page, four-color spread is a decadent piece of 1970s advertising, the scene's neutral, wood grain tones punched through with vibrant yellows, oranges, and greens. We join a couple in the kitchen. The interior suggests an open floor plan home in the bungalow style so common to the California dream; through the expansive kitchen window, a lush backyard. In the middle ground, a woman chops produce on a woodblock cutting board. Around her are the elements of domestic life, carefully chosen: a tall white teapot, a metal mixing bowl, a crisp white KitchenAid, a bowl likely full of fruit, a toaster oven perhaps. Behind her, a white framed photograph of a Red Delicious apple, a coy bit of set dressing.[2]

But follow her gaze. In the foreground is the working modern man: a pen and a cup of coffee, a casual blue turtleneck, an exposed watch, no wedding ring. His eyes rest on his fingers, curved over the keys of an Apple II. Just to the right, behind a stack of papers, a color monitor displays the quantification of . . . something. Orange is up; pink is down; green is steady but minimal. The scene is devoid of the complex apparatus of computing's reality: the cables and cords, the cassette deck, the tapes and cases, the user manuals. Nonetheless, the computer is in the *kitchen*, not tucked away in the den or library. The man works in the company of his mate. They are a part of things, altogether, changed by technology but somehow still fulfilling their expected roles. There's nothing science fiction here. This was the future Apple was selling: a present you were already living but more.

Seven years later, however, American households weren't biting. While the 1980s opened with industry projections that computers would quickly become ubiquitous in American households, it was estimated that the installed base of computers in the home in 1983 was only around 6 million, in a country of approximately 84 million households.[3] Despite the advertisement's claims that learning BASIC would empower Apple II owners to "store data on household finances, income tax, recipes, and record collections" and despite a market of microcomputers full of options and price points, most consumers had yet to be sold on the idea of a computer at home.

Apple's 1977 advertisement for the Apple II had to lean on the versatility of BASIC for the at-home consumer as the software market at the time was nearly nonexistent. Yet over the coming years, an unruly hodgepodge of products came to be organized under categories like "home," "home productivity," "home management," or even just "personal." Compared to the other software discussed thus far, "home" software would embody the greatest range of applications and functionalities, spanning tools for the digitization of household references and management of lightweight financial tasks, wide-ranging exercises for tracking physical health data, self-help amusements, and computational novelties that skirted the edge of games. Few of these products would ever sell in sufficient volume to dominate best-seller lists, but that isn't the point, historically speaking. In this breadth of applications for the home, we find cases for computer use that are deeply personal, suggesting the emergence of a new range of technological intimacies that early adopters of microcomputers were eager to explore, even if only to experience how using a computer might reorient their daily lived experience.

Yet the home software market struggled to find its hits. As Dan Gutman diagnosed it in 1984, mainstream software publishers were almost entirely occupied with trying to devise a "VisiCalc for the home, one magical program that will sell the computer to the home market the same way VisiCalc sold it to the business market."[4] Thus far, in his estimation, that market impulse had largely resulted in an unimpressive sea of spreadsheet and word processing copycats—"Thiscalc, Thatcalc, Wordword, Nerdword," as he derisively chimed.[5] Gutman's solution was that mainstream software publishers needed to imagine use cases as idiosyncratic as users themselves, software to "help him plant his garden, chart his biorhythms, plan the route of the family vacation, fix his car, choose a college, or pick a career." Yet given how small the market was, Gutman's was an uphill battle: How could you convince a mainstream publisher to offer software for such a small subsegment of the user base? How could you create software extremely specific to the person who used it but simultaneously universal enough that everyone would buy it?

Enter *The Print Shop*, developed by David Balsam and Martin Kahn of Pixellite Software and published by Brøderbund Software in 1984. *The Print Shop* was a menu-driven program for laying out text and graphics, which, when used in conjunction with a printer, allowed users to create banners, signs, greeting cards, and other simple printed materials. *The Print Shop* trucked in none of the traditional rationales for home software: it digitized no knowledge, created no efficiencies, and quantified nothing. Instead, the appeal was in the sheer novelty of customization, of endlessly combining borders, fonts, text layout, and graphics into personalized mementos or forms of public communication. Rather than feel constrained by *The Print Shop*'s limited design capacities, users found delight in what was perceived as a computer-aided *expansion* of their personal creativity. In *The Print Shop*, we find a version of home computing that reflects broader shifts in everyday computer practices, as users sought ways for their computer experiences to exceed the limits of screens and disks.

Title screen from *The Print Shop*, running in MAME emulation on the Internet Archive (top), and a printed greeting card created using the program (bottom). Screen capture courtesy the author, photo courtesy Tega Brain.

The Print Shop was also published at a moment when the shape of the entire "home" computing category was undergoing a distinct transformation from where it had been just a few years before. Thus we must move *through* that history in order to appropriately situate the program's development, publication, and consumption. What follows will trace how the changing character of microcomputer users can be tracked through shifting notions of what constituted "home" software before *The Print Shop*'s release. By 1984, hobbyist and professional users were no longer seen as the only market for microcomputers. Hobbyist notions of computing at home—which often focused on computerization *of* the home, or simply the advancement of computing as an at-home hobby—were overtaken by a more encompassing, marketable concept of "home computing," geared to a less technical audience and fueled by the economic investments of a growing software industry. Tremendous effort went into circulating and upholding the notion that, as one mid-1980s buyer's guide put it, "computers [are] for everybody."[6]

The Print Shop reflected these concerns at the level of software design and conceptualization. Its menu-driven interface was so easy to use that the software was routinely lauded for not even requiring the user to read the manual. Even its developers, Balsam and Kahn, expressed they were "liberating people's artistic imaginations" by not requiring them to create the images or underlying compositional structures themselves.[7] *The Print Shop* thus displaced experimentation with the computer hardware or programming in favor of experimentation with the limits and capabilities of a piece of consumer software. In doing so, it modeled the kind of consumptive creativity that would become a hallmark of computer use among a slowly rising tide of nonhobbyists.

* * *

As noted in chapter 1, computing's traditional inaccessibility to a consumer population was a direct result of both the cost and the size of computing components. General-purpose computing devices were nearly nonexistent in homes before the mid-1970s.[8] While a few stray examples exist in the historical record—most notably, ECHO IV, a bespoke computing system built by a Westinghouse engineer in the mid-1960s, and Neiman Marcus's 1969 Honeywell Kitchen Computer—such systems unanimously prove the rule.[9] While these systems could perform some of the claimed computational activity, neither was particularly practical or useful. ECHO IV and the Kitchen Computer existed as popular oddities rather than prototype technologies. More telling is the fact that media coverage of both systems attempted to neutralize the threatening or disruptive potential of computing through appeals to conservative notions of the family with a gendered division of labor—a theme that would amplify, as we shall see, in the early to mid-1980s as marketers sought ways to sell US consumers on the benefits of personal computing.

For electronics hobbyists in the 1970s, however, the dream of a computer you could own necessarily meant a computer you could bring home. This notion propelled the editorial accompanying the January 1975 issue of *Popular Electronics* that featured the Altair, with Art Salsberg, editorial director, writing, "For many years, we've been reading and hearing about how computers will one day be a household item."[10] The Altair was perceived as an answer to that promise: a way for hobbyists to get their hands on computing power outside of the offices, science labs, and research centers that had been computing's natural habitat for decades.

Ambitions regarding the purposes a microcomputer might serve, however, were tempered by the technical limitations of such machines; discussions of microcomputer applications for home or personal use remained largely aspirational when they existed at all. A survey of the first two years of *BYTE* returns minimal examples.[11] In terms of executable code listings, only one notable example exists: a 1976 BASIC program for computing biorhythms, a popular physiological pseudoscience of the 1970s.[12] While the limitations of first-wave microcomputers like the Altair were significant, the authors and readers of hobbyist magazines like *Popular Electronics* and *BYTE* understood themselves as working toward a future in which such applications *would* be more plausible, a sentiment expressed by one author who hoped his flowchart for a completely impractical kitchen database system would "help catalyze development efforts in what appears to be a fruitful home computer applications area."[13] Yet in reality, most hobbyists were more occupied with the struggle of implementing, building, or even just getting their hands on a machine like the Altair than they were with figuring out exactly what they might do with it.

Even after the release of the 1977 Trinity, Apple's ambitious kitchen ad, and other more "user-friendly" microcomputers, the idea of a computer in the home remained aspirational if not outrightly fantastical. Consider, for example, the cover of *BYTE*'s January 1980 issue dedicated to "Domesticated Computers" (see pl. 14). The cover illustration is an unusual piece of elitist future-gazing: foregrounded by small white gloves, a string of pearls, and a champagne coupe, the computer, encased in a carved wood cabinet, formally announces, "MADAM: DINNER IS SERVED," on its CRT screen. Here the computer is a servant to the lady of the house, seemingly managed by an antennaed remote control—the image's only nod to the boxy tech stylings of the late 1970s. The scene is more than merely domestic; it conveys deep associations with hereditary white wealth, situating these supposedly future-oriented technologies in a retrograde fantasy of masters and servants. And the propositional narrative is curious: Did the computer make dinner, or is it only relaying the cook's message? What, exactly, does the remote control *control*? This is the first *BYTE* cover in which the implied user is a woman, but the encased computer, the calculator-like remote, and the cocktail-hour set décor imply also the feminized user kept at a firm distance from the technical operations of all equipment involved. This is push-button domestication, a vision of computing in which technologies are *used* but certainly not *worked on*.

The aspirational cover is at odds with the applications detailed in the issue, however, exemplifying the continued tension between computer fantasies and computer realities. The issue features four articles about computerized home applications, all focused on systems control of appliances and utilities, including lighting, furnace management, and automated phone dialing. This isn't a world in which domestic labor is productively delegated to the computer and therefore disappears. Rather, it showcases how the computer multiplies domestic labor by creating information networks across domestic space where none had existed before, each requiring its own management, control, and regular maintenance (to say nothing of the need for an expensive computer solely dedicated to the task). The opening vignette from the issue's article "Computerize a Home" precisely illustrates this, as the article's author, Steve Ciarcia, details his efforts to run the household lighting system from a computer in the basement. After accidentally turning various upstairs room lights on and off, the author's wife, Joyce, calls down to her husband in the basement to ask if a fuse has been blown:

> I grinned in a way that only a Cheshire cat could appreciate. "Sorry, Joyce, just experimenting on the latest article." Chuckling softly, I continued. "I hope you don't mind, but the computer seems to have taken over."
>
> "Can it make beds?" she replied.
>
> I should have known that she wouldn't be taken in so easily. "OK, I'll tell the computer to keep its sphere of influence to the cellar."[14]

Joyce embodies the trope of the patient but unimpressed wife, waiting for her husband to make the computer do something useful. Not only is the computer decidedly *not* a machine taking over material household work; it creates new responsibilities in the process of domestic integration. In this issue of *BYTE*, "domesticating" the computer is mostly a sleight of hand in which *inventing tasks* masquerades as *simplifying lives*.

The hobbyist era of microcomputing, stretching from the mid-1970s to the dawn of the 1980s, thus demonstrates the considerable effort that went into imagining uses for the computer in domestic space. This was a project that required continual reinvestment, in which fantastic hopes for what a computer might one day do were an a priori pardon for the limited, and often labor-multiplying, results of the present. The promise of a computer that might do more, dreamed of by hobbyists and continually leveraged as "almost there" by magazine editors and marketers, was the fuel that chugged these fantasies forward—and that would eventually be catalyzed in very different ways to appeal to an expanded range of nontechnical consumers.

* * *

While software developers had dabbled in applications for the home since the first wave of commercial microcomputers, the uptake of "home" or "personal" software was slower than that of other categories, like business or games. *Softalk*'s first best-seller list, from October 1980, featured only three pieces of software that could plausibly be categorized for home or personal use—Microsoft's *Typing Tutor* and two word processors—and even these were not exclusive to domestic or personal use.[15] As the only category with no significant precedent in earlier professional computing cultures, there was a unique complexity to identifying what "home" software might be. More readily defined by what it wasn't than what it was, "home" software is only an index offering us clues to a much stickier set of historical circumstances: What were people even doing with computers in their home?

As the earliest magazine to offer a comprehensive best-seller list for Apple II software, *Softalk* provides some traction for understanding how the broader community of retailers, journalists, and enthusiasts understood these emerging categorizations, which shifted over a remarkably short period as the software market and the user base grew together. In an effort to give readers, as well as developers and publishers, more granular information on the growing software market, *Softalk* began running a supplementary "Home/Hobby" top ten list in February 1981, just five months after its launch issue.[16] The "Home/Hobby" category primarily comprised assemblers, disk duplication software, and graphics packages; the "home" component of software sales on this chart included only *Typing Tutor* and *Dow Jones News & Quotes Reporter* (word processors had already been refiled as "Business" software). This alignment of "home" and "hobby" software types reflects an assumption that the home computer user was a serious hobbyist, someone whose passion for computing led them to deploy the computer for various niche and idiosyncratic household purposes, such as receiving news at home via a modem. By May 1981, *Typing Tutor* had become the top seller in this category, while Continental Software's *Home Money Minder* and Howardsoft's *Tax Preparer* both moved into the Home/Hobby top ten. Other non-utilities appearing on the Home/Hobby top ten over the next few months included *VisiTerm*, a modem-based file transfer communications tool, and *"The World's Greatest Blackjack Program,"* which was not categorized as a game because it taught strategy instead of just serving as a simulation. Yet with the exception of *Typing Tutor*, none of this motley assortment of home/personal software ever broke into *Softalk*'s broader top thirty, indicating that much of this software was still uncompetitive compared to business applications, games, and a few top-performing utilities.[17]

In the October 1981 issue, *Softalk* revised its best-seller breakouts to "better reflect the diversity of special interests within the Apple community,"[18] adding top five categories for word processors and several individual game genres and separating "Home" and "Hobby" into distinct lists. Reflecting on their decision, the *Softalk* team wrote:

> As to the new breakouts, they still cause some problems in categorizing software. Some of the divisions were fairly clear, but the software doesn't necessarily want to fit the niches defined.
>
> As an example, it seemed natural to break out the Home/Hobby 10 into a Home section and a Hobby section, with the hobbyist being considered a person who does programming. But do Sensible Software's utilities fall into that category, or into the general home use category? Likewise, are Graphtrix and Hand Holding Basic the tools of hardcore hobbyists or are they more home-oriented?

> As with *Softalk*'s original breakouts, the divisions were made arbitrarily and are susceptible to knowledgeable second-guessing by one and all.[19]

Self-deprecation aside, *Softalk*'s iterative approach to software categorization (it altered its strategies twice in a year) is a testament to the rapidly changing character of Apple II users. The hard-core hobbyist was no longer *the* assumed user of the Apple II but one of several types of user, including gaming enthusiasts, business users, and home or recreational users. Similar concerns about "the broader market" of word processors drove *Softalk* to separate text handling software from the "Business" category, recognizing "many home users buy *Apple Writer*, but few buy a $500 general ledger accounting package."[20] "Home" software might have been an amalgam of everything from typing instruction software to home finance packages to beginners' programs for learning introductory BASIC to telecommunications services, yet what held much of it together is the way the software, as a whole, configures the microcomputer as a tool for *extending* one's capacities in a nonoccupational setting, encompassing users with hobbyist or professional computing competencies while also reaching beyond them. While a person might span multiple use categories (most business software sold with at least a couple of games, and a hobbyist might own a home finance management system as well as a compiler or assembler), these emerging divisions reflect an excited awareness within the industry that the Apple II was exceeding the important but narrow emphasis on hobbyist or serious business users.

These observations align with broader trends in the shifting character of microcomputer users, especially for the Apple II. Previous research by me and the computing historians Kevin Driscoll and Kera Allen has demonstrated that 1980–84 marked a transition from "programming to products," in which "microcomputing's use cases . . . shifted away from hobbyism."[21] Our analysis of over twelve hundred letters to the editor published in *Softalk* revealed a progressive shift toward less technical topics, which evidences "the expanding constituency of those we would identify as *users*: microcomputer owners whose investments were not located in the microcomputer itself but in what it might do for them."[22] While this trend toward greater accessibility was common across nearly all software categories, it was especially salient in the home software market, where preexisting experience with the computer could not be guaranteed, and there was little occupational pressure to computerize one's life. For some users in the mid-1980s, software consumption itself constituted a new hobby, in which the focus was not the technical arcana of the computer but the "pleasure [taken] in the expert application of microcomputing to everyday life."[23]

In terms of the development of the home software industry during this time, a few important qualities distinguish this category from others discussed in this book. First, despite the wide-ranging array of software designed for home and personal use, the best-selling software in this category was largely restricted to typing instruction software, home finance packages, telecommunications programs, and a few general-purpose (i.e., nonbusiness) word processors. These software subcategories reflect the most generic personal applications, and popular software of this sort tended to have significant incumbency on the best-seller list. Programs like Brøderbund's *Bank Street Writer*, Apple's *Apple Writer*, Microsoft's *Typing Tutor*, Lightning Software's *MasterType*, and Continental Software's *Home Accountant* all maintained top thirty best-seller list standing for a year or more, showing sales consistency on par with major business applications. Yet while the most popular programs were typically manufactured by well-capitalized companies embedded in a small handful of major software distribution networks, these same companies typically avoided delving too deeply into more specialized home and personal applications. As the Sierra On-Line cofounder Ken Williams opined in the pages of *Creative Computing* in 1984:

> It is expensive to develop software. All software must be developed to reach the widest possible market. If your software applies only to men or only to women, you have already lost half of your potential customers. . . . As to even more narrow applications like help in completing

Boy Scout projects or painting your house, forget it. The market is too small.[24]

For Williams and other major publishers, the governing wisdom was that highly individualized applications made for limited markets. Small product runs weren't worth the time or investment, especially as the market grew toward early maturity and software publishers took on venture capital in exchange for the promise of dramatic returns. Consequently, the development of more niche programs was left to bottom- and middle-tier firms that sold by mail order through word of mouth or small magazine ads. Thus while best-seller lists like those in *Softalk* offer a good overview of what software was most popular among the general Apple II user base, they are inadequate for grasping the full breadth of the category—a breadth that is essential for understanding exactly what was being promised by the home computing "revolution" at large.

* * *

The business of selling computers in the late 1970s and early 1980s had relied on discursive appeals to hobbyists who had already bought into microcomputing's potential. These appeals changed by the mid-1980s as the industry sought to sell computers to the masses. Early hobbyist fascinations with computerized homes continued but were also expanded in order to reach a broader class of wealthy Americans—a turn from "computing at home" to "home computing," as it were. In this sense, home computing was more than simply a software category; it embodied an ethos of living with computers, of imagining one's *life* (and not just one's home) as available to be computerized. Importantly, such efforts were disconnected from the countercultural ambitions often associated with early homebrew computing culture, like those embodied in the futurist Ted Nelson's routinely circulated proclamations about "computer liberation" or the Hacker Ethic.[25] Rather than breed a new generation of enthusiasts empowered to achieve greater freedom through the personal computer, these new discourses of home computing largely focused on normalizing the computer as a necessary but unobtrusive partner in middle- and upper-class life.

No singular event or artifact affirms this shift. Rather, it accumulates in a deluge of mundane primary documents that begin to emerge from 1983 on, becoming particularly apparent when one examines the longer arc of computing enthusiast magazines during this time. Magazines founded in the mid- to late 1970s—including *BYTE, MICRO: The 6502 Journal*, *Kilobaud*, *Nibble*, and *COMPUTE!*—approached microcomputing as a firmly technical leisure activity in keeping with the hobbyist user base. The content of these magazines reflected this, emphasizing program listings, schematics, intensive hardware reviews, and a fairly high threshold for comprehension. *Softalk* was one of the first magazines to buck this standard in September 1980, declaring itself "not a programming magazine" in the opening issue; its cover stories focused on human-interest content, and its beginners' resources aimed to aid those curious but unfamiliar with the finer points of programming.[26] This editorial turn toward accessibility was followed by *Popular Computing* in 1981, which aimed to "demythologize small computers in a direct and entertaining manner."[27] This trend amplified into the mid-1980s with magazines like *A+*, which focused on delivering approachable content to consumers who used an Apple II in professional settings, and *Family Computing*, dedicated to "[helping] families in the years ahead . . . feel comfortable as casual or recreational computer users."[28] *A+* and *Family Computing*, which both launched in 1983 with women as their editors in chief, embraced a softer, less technical tone than their predecessors and were designed more like lifestyle magazines than technical journals, deploying a more nuanced typographic style and a greater reliance on colorful infographics, photography, and illustrations. Importantly, *Popular Computing*, *A+*, and *Family Computing* were all launched by major publishing conglomerates (McGraw-Hill, Ziff-Davis, and Scholastic, respectively), indicating the extent to which the mainstream media had a vested interest in tapping a larger market of potential microcomputer owners.

Just as *BYTE* consolidates the spirit of the hobbyist approach to microcomputing, *Family Computing* proves instructive on how "micros" were being reframed as "personal computers" for consumption within the domestic unit. The cover of the magazine's September 1983 launch issue provides its own take on the notion of domesticated computing (see pl. 15), offering a productive parallel to both Apple's original 1977 Apple II ad and *BYTE*'s 1980 domesticated computing issue. While the theme of computer-as-domestic-possession still dominates, *Family Computing*'s approach is greatly humanized compared to both previous examples. Gone is the master/servant dichotomy and the emphasis on command-and-control that dominates *BYTE*, the fixation on the computer as a stand-in for upper-class power relations. And whereas Apple's ad center the microcomputer as a domestic appliance, a tool for collapsing work and home, *Family Computing* situates the computer as *part of* the family, awkwardly seated on the lap of the elder son like a family pet. If *BYTE*'s cover featured the computer as a tool for sustaining racialized class relations, *Family Computing* displays a distinct but not incompatible fixation on maintaining the status quo in the form of the white, heteronormative, upper-middle-class family. The computer is portrayed as an inheritance of a different sort—not an expression of wealth that can be controlled, but something that would live and grow with the children as part of the perpetuation of intergenerational privilege. The message is clear: the computer is safe; the computer is part of your team; the computer isn't just present but *fits in*.

The impulse to represent computers as "fitting in" to domestic life hints at more complicated dilemmas. Those intent on selling computers to the masses had to negotiate popular anxieties about computers while demonstrating that the computer was useful and accessible. As the media scholar Lori Reed has shown, computerphobia was a documented phenomenon among the American public dating back to the mainframe era, with ample evidence suggesting that "people resisted computer technology very strongly" well into the 1990s.[29] The notion of computerphobia was prevalent enough to appear in psychological manuals in the mid-1980s, and the existence of books like *Overcom-*

ing Computer Fear and *Technostress*, both published in 1984, indicates that "for many people, computers were extremely unpleasant objects and their introduction into some people's lives caused extreme duress."[30] However prevalent among hobbyists and futurists, discourses about the home computer revolution were largely lost on the broader American public.

Instead, negotiating consumer anxiety about computers required starting from the premise that people could overcome their computerphobia through education and exposure, as illustrated in the early 1980s beginners' guide *Computers for Everybody* (a book so popular it had several editions and was reprinted in 1983 as a Signet New American Library paperback). In the book's first chapter, "Computers Aren't Scary Anymore," the authors, Jerry Willis and Merl Miller, specifically address the computerphobic consumer:

> One of the biggest problems most people face in dealing with computers is fear. We've been taught to think of computers as big, complicated machines that, in the wrong hands, can do terrible things. We've been told for many years that the reason the bill was wrong was because the computer made a mistake. In reality, what happened was a human made a mistake and blamed it on the computer.
>
> The computer is just like any other machine or tool. It is only as good as the person who runs it and only as useful as you make it. You can choose to use it in any manner you want. It won't take over your life, run off with your spouse, or cause you strife.[31]

Such repositioning of the computer—from a machine out of your control to a tool under your control—was a common editorial approach to assuaging dubious and anxious consumers. The first issue of *Family Computing* featured an extensive essay titled "Confessions of a Reformed Computer Phobic," which focused on the trials and challenges of a middle-aged male Apple II owner who overcame his fear and sense of isolation to master the computer on his own terms. The article included a sidebar, "RX for Computer Phobia," suggesting new users "choose something that's fun to do" and "keep your expectations within realistic bounds."[32] The repeated message for the computerphobic was that diligence, patience, and self-compassion could transform the computer fearing into the computer loving.

But "fitting in" was not just about allaying fear. The computer also had to be perceived as useful, which wasn't always immediately apparent. Potential users had to be taught how to imagine their lives as available for intervention by a computer, instructed on the ways in which computing could replace or augment other kinds of tasks.[33] Sometimes these efforts took on a straightforward form, with computer enthusiast articles coaching stay-at-home moms on how a computer could let them run a business from home or suggesting which software was best for managing and tracking gardening tasks, but more creative approaches were also common.[34] One recurring editorial trope was to feature stories based on a diary or day-in-the-life format, which allowed readers to map the use of a computer within the home over a period that ranged from days to months.[35] Such articles often started by acknowledging some anxiety on the part of specific (often female) family members that resolves over time. For example, Joan Levine, the wife and mother documented in "A Family Computer Diary," starts out as a "self-confessed computer-phobe" but is using the computer to compose her dissertation nearly a year later.[36] Similarly, Robin Raskin's "Let's Be Friends: The Diary of a Family's First Day with Its Computer," which appeared in *Family Computing*'s January 1984 issue, is a utopian portrait of how a family of six works together to unpack, connect, boot, and use their first computer. One child reads the manual, another opens the boxes, Dad fills out the warranty card, and Mom has already invented a computer "chore wheel" for handling tasks like cleaning the screen, inventorying printer paper, and making sure disks stay organized. Such features were a pivot on the kind of human-interest content pioneered by magazines like *Softalk*. But whereas *Softalk* focused on the use of the Apple II in unusual or captivating settings—from handling special effects for *Star Wars* to tracking plant growth in outer space—these kinds of articles repositioned the computer as a day-to-day appliance that could transform domestic life into a high-tech yet family-friendly experience.

Another tactic employed to convey the usefulness of computers was itemizing the potential applications for computing in the home—the various ways one's life could be computerized. As the family in "Let's Be Friends" moves on to actually using the computer, for example, each family member is encouraged to find a purpose appropriate to their needs: the children play recreational and educational games, while the adults balance family budgets, store recipes, and type letters. *Computers for Everybody* similarly illustrates this trend by breaking down the functionality of home computing into six areas that reflected the broad scope of home applications:[37]

- Entertainment
- Personal development
- Personal finance and record keeping
- Hobby and recreational programs
- Home control
- Computers and health

Developing these sorts of subcategorizations fit well into the larger project of helping consumers imagine how computers might map onto their personal lives. If the computer was, as Willis and Miller suggest, "like an artist's blank canvas," wherein "the outcome will depend on the artist or the user," these categories served as a conceptual palette that helped consumers sort through a vast and confusing array of home software.[38] By framing software applications through use patterns, consumers were encouraged to imagine many ways the computer could meet their already ambient "personal" needs.

As we can see, this new breed of magazines and beginners' guides went to great lengths to emphasize how computers could respond to the unique and individual contours of each user's lived experience. Distinct from the hobbyist ethos of magazines like *BYTE*, which circulated applications and tutorials to advance enthusiasts' sense of technical discovery of the computer itself, this revised discursive approach to "home computing" encouraged a new generation of users to see their lives as the raw material for expansion *through* the computer. Yet despite the phenomenal amount of money and effort invested in positioning the microcomputer as a domestic technology, such techniques were not without their limits. As highlighted in this chapter's introduction, the notion of home computer software still had yet to prove its value for the majority of users. It was one thing to convince people to buy a microcomputer for occasional word processing or digitized checkbook balancing; it was quite another to turn them into recurring purchasers of software. The solution would not come, however, from the vast proliferation of niche uses. There *would* be something like a "VisiCalc for the home." It just came in a form no one was expecting.

* * *

If the alleged promise of home computing was the articulation of an expanded self, then making good on that promise required the deployment of software to support such presumably "expansive" purposes. Home budgeting software might allow one to locate financial inefficiencies. Typing and programming instructional software could bestow new skills on its users. Word processing could remove the drudgery from traditional typing, from school papers to personal correspondence. And rudimentary calorie counters or fitness trackers might offer new knowledge of the body. But being *creative* with the computer wasn't so easily done. *Making things* was a challenge, if not an impossibility, for the everyday nonprogrammer, and the fact that things made on computers had to stay on computers limited the extent to which one's creations could enter the broader world.

Keep all this in mind, and the outstanding success of a program like *The Print Shop* makes a lot more sense. While the capacities of *The Print Shop* may seem parochial to the twenty-first-century computer user—after all, what did the software do besides print out a little bit of text with an image or two and maybe a border?—it was perhaps the first piece of software that resolved the creative accessibility gap in the promise of home computing. In other words, *The Print Shop* offered a constrained but well-designed environment that allowed users to combine text, images, and form in ways that were not just personally meaningful but also had a physical, noncomputational existence. You didn't just "make things" in *The Print Shop*; you printed them out and showed them off.

Yet *The Print Shop*'s journey from ideation to execution was far from a straight line. By 1984, the year of the program's release, the software market was highly congested, and there had never been more users who knew less about computing. Long gone were the days, just a few years back, when the market was "very, very tolerant of very obscure and complex procedures" or when a single self-published piece of software could launch a new business.[39] New software had to cut through a tremendous amount of noise. *The Print Shop*'s iterative development history, as we shall see, precisely illustrates the new thresholds of polish, appeal, and accessibility software had to meet in order to stand a chance of thriving in an overcrowded yet often underwhelming market.

The Print Shop was the outgrowth of a creative and business collaboration between David Balsam and Martin Kahn, a gay couple in their late twenties who met in the San Francisco area in the early 1980s (see pl. 16). Both had relocated to the Bay Area in the 1970s, drawn to explore their nascent identity in what was then a vibrant locus for the gay liberation movement in the United States. Balsam had come from New York City, where he had attended the experimental John Dewey High School in the early 1970s and reaped the benefit of access to the school's computer lab. Though he did not pursue programming or computer science as an immediate occupation (he took a job copyediting the Yellow Pages in Manhattan to save money for the cross-country trip in 1977), he had tinkered enough with the systems to be able to program a simple game in BASIC—granting him a familiarity with computing that would resurface when he met Kahn.[40] Kahn, for his part, had been raised in Los Angeles and had transferred from UCLA to Berkeley for his last two years of college. He graduated in 1976 with a double major in linguistics and mathematics, subjects he focused on because they felt "easy" to him. He had taken enough programming classes to get himself a job in the computing industry immediately out of college, eventually landing at NorthStar Computers in the late 1970s, where he developed graphics utilities for the company's business-facing microcomputer, the NorthStar Advantage.[41] While at NorthStar, Kahn used his downtime and access to company facilities to create computer-generated art, visually rendering mathematical equations as complex volumetric forms, printing them out, and sometimes painting them.[42] Balsam was captivated by Kahn's work, which covered the walls and more than a few windows of Kahn's apartment: "When I first saw the paintings . . . they had the quality of an Escher or a Vaseralay [*sic*], three-dimensional shapes moving behind other shapes, distorted perspectives, fascinating topological landscapes."[43]

The romantic relationship between Balsam and Kahn was complemented by a burgeoning technical and creative synergy between the two men. Remembering his high school fascination with computers, Balsam took to playing with the NorthStar computer Kahn kept at home and eventually quit his job working for an insurance company to become a salesman at a local microcomputer retailer, selling computers to small businesses. Meanwhile, Kahn grew restless with his work at NorthStar ("They didn't give me enough to do," Kahn sardonically recalls of his time there).[44] When Balsam showed him a classified ad soliciting programmers to work at a small, local entertainment software company called Brøderbund, Kahn applied. Kahn's low-level programming experience with the Advantage's Z80A microprocessor was more than sufficient to secure an interview and employment; even though the majority of Brøderbund's products were targeted to 6502 platforms such as the Apple II, the Commodore 64, and the Atari 8-bit family, knowing *how* to work in assembly language was a desirable skill. Kahn started in January 1983 and was immediately thrown into programming the Commodore 64 port of the popular Apple II pinball game *David's Midnight Magic*.[45]

The energy of Brøderbund was youthful and electric. As Balsam recalls, Kahn went from working at an "innovative but stodgy

Two samples of computer art by Martin Kahn, 1981. Both images were created on the NorthStar Advantage using mathematical formulas written in BASIC and 8080 assembly language. Printed on a daisywheel printer, using only the "." (period) key. Printed resolution is 640 × 240. Courtesy Martin Kahn.

business-oriented company to a very young and very dynamic, exciting environment with lots of young programmers."[46] When Kahn was hired, Brøderbund had only been in San Rafael for about fifteen months, having relocated from Eugene, Oregon, in September 1981. The company's emergence was typical among very early entrants in the entertainment software business: it began as a serious hobby that grew into a serious business.[47] At the center of it was an Ivy League lawyer turned "electronic-age vagabond" named Doug Carlston, who drove around the country in 1979 selling cassettes and floppy disks of a TRS-80 space colonization game he had programmed called *Galactic Empire*.[48] His road trip eventually ended at the door of his brother, Gary, who was eking out a living in Eugene. On a lark, Gary tried selling some of Doug's software himself; one $300 order later, they determined they were a business and formally registered a company together in February 1980. The name they chose, "Brøderbund," was a bit of pig Scandinavian, a mash-up of Swedish and Danish and Dutch that gestured, in loose translation, to an "alliance of brothers."[49]

The workplace Kahn walked into three years later had grown from the two Carlston brothers (who were joined by their sister Cathy early in 1981) to roughly thirty-five employees. By the end of 1982, Brøderbund was a middleweight titan of independent software production, ranking third against its competitors, Sirius Software and Sierra On-Line in overall consumer sales for the year.[50] While most of that product was games, and most of it was created for the Apple II, Brøderbund was, like its competitors, rapidly expanding into other microcomputers as well as more "serious" software categories. The company's first general-purpose word processor, *Bank Street Writer*, had been released in 1982, growing into one of Brøderbund's best-selling products in 1983.

But Brøderbund was also unusual for operating almost exclusively as a *publisher* in the entertainment software space rather than trying to professionally publish software while also nurturing internal talent (like Sirius Software did with Nasir Gebelli) or funding programmers who were still early in development (as Ken Williams was prone to doing at Sierra On-Line). Software like Tony Suzuki's *Apple Galaxian*,

Dan Gorlin's *Choplifter*, Douglas Smith's *Loderunner*, and *Bank Street Writer*—products that were not just successful for Brøderbund, but were some of the best-selling and most widely beloved releases of the early 1980s—had all come in as polished prototypes or completed programs.[51] In other words, Brøderbund was a company that knew what it was good at, namely, identifying original, promising products with potential widespread appeal and setting them up for successful release. Evaluating outside software was such a significant activity that it was even embedded in the company culture: Friday evenings were sometimes given over to the "dog and pony show," as Balsam called it, a rowdy collective event in which Brøderbund management demonstrated recent submissions and let employees have their say about what should make the cut.[52]

It was in the context of this porous, collegial atmosphere—where management made a habit of letting employees weigh in on software submissions, where no one made much mention when a male programmer brought his boyfriend to the Friday night company hangout, and which was nestled in the larger context of the Bay Area, where everyone seemed to be striking it rich in software—that Kahn and Balsam wondered how they too might get their piece of the pie. As they brainstormed what they might create, the pair focused on two key points of leverage. First, they wanted to make good use of Kahn's graphics experience, giving users a way to make their own images and artwork without the tedium of using complicated professional tools. Second, they were interested in devising a way to share these creations *between* users.

Following these inclinations, Balsam and Kahn developed a prototype called "Perfect Occasion," which Balsam described as a "greeting disk sort of a thing."[53] The idea was to give users basic tools to create their own images or animations and superimpose text on those images to create a special, personalized message—which could then be shared with the recipient via floppy disk. "We thought it would be fun to make something like that," Balsam stated in a 1985 interview with *Microtimes*, "so you could send it to a friend and they could put it on their computer and be amazed by it."[54]

Despite Perfect Occasion's novelty, however, Brøderbund wasn't ready to bite.[55] The general response was that the concept looked fun but was limited in scope. As Balsam later put it, "Not everybody's grandmother had an Apple then, and seriously, how many Valentines do you send to members of your user's group?"[56] Yet something interesting was going on here. If much of what underlay consumer computerphobia was the computer's murky association with automation and human disempowerment, Perfect Occasion suggested that a computer-generated object could operate as a unique and effective mode of *human* expression. In thinking through how the software might be redesigned to capture a larger market, a member of Brøderbund's product development team offered a curious piece of feedback: What if, aside from creating digital greeting cards to share, you could create physical ones? In other words, what if it could print?

* * *

"What if it could print?" turned out to be a prescient question. While printers were not exactly ubiquitous among home computing users of any platform, the Apple II included, they were gaining popularity. Falling printer prices, alongside an expanding market of versatile word processors and small business applications, increasingly weighted the cost-benefit analysis in favor of printer ownership. In 1983, *Creative Computing*, *Softalk*, and *A+* all ran features on how to choose a printer, positioning these peripherals as the next step in mastering the computing experience.[57] As *A+* put it to its readership, printers were necessary because they made the benefits of computer use visible to others:

> By itself, the Apple is an actor without an audience. Although it can tell tales of dismay and delight about the future using mathematics and modeling, it keeps them a secret, privy to the select few who can see its monitor screen. . . . A printer can change all that. A contrivance that's half electronic brain, half mechanical beast[,] . . . [t]he printer

frees your computer's hidden secrets and makes them available to the rest of the world.[58]

Thus, while printers were not perceived as essential accessories in the mid- to late 1970s (especially among home microcomputer users), they were discursively repositioned, by the mid-1980s, as a core technology for not just getting the most *out of* one's computer, but *living up to* the potential of computational life. Establishing these kinds of desire trails did not become possible, of course, until printers hit the edge of upper-middle-class affordability. Between the late 1970s and the mid-1980s, the cost of printers dropped dramatically. *Creative Computing* reported that by 1983 there were roughly twenty-five printers available for under $1,000, as opposed to only two in that price range in 1979.[59] Apple Computer even offered a variety of branded printers during the early to mid-1980s—most prominently, the Apple Dot Matrix Printer (1982) and the ImageWriter series (1983 and 1985)—all of which retailed for $600 to $700.[60]

Back at the drawing board, Balsam and Kahn worked on reconceiving the greeting card concept in print form—a puzzle that did not have an immediately obvious solution, given the constraints of consumer-grade printers. As physical objects, greeting cards use both sides of a single piece of paper: the outside of the card, which includes the front and back covers, and the inside of the card, which is traditionally where the card giver writes a personal message to the card recipient. This design requirement, however, did not translate easily to printers of the 1980s, which could not do automated duplexing, or double-sided printing, because of hardware and operational constraints. Consumer-grade printers required paper to be manually loaded using a spooling mechanism, and printer paper itself was typically sold in long, perforated sheets that had to be separated by hand once printing was complete. While it certainly would have been possible to print on one side of a sheet of paper, tear that sheet off, then load it backward and upside down back into the printer, such a process would have been finicky and prone to error—not the kind of challenge that would have made the software fun or easy for the user.

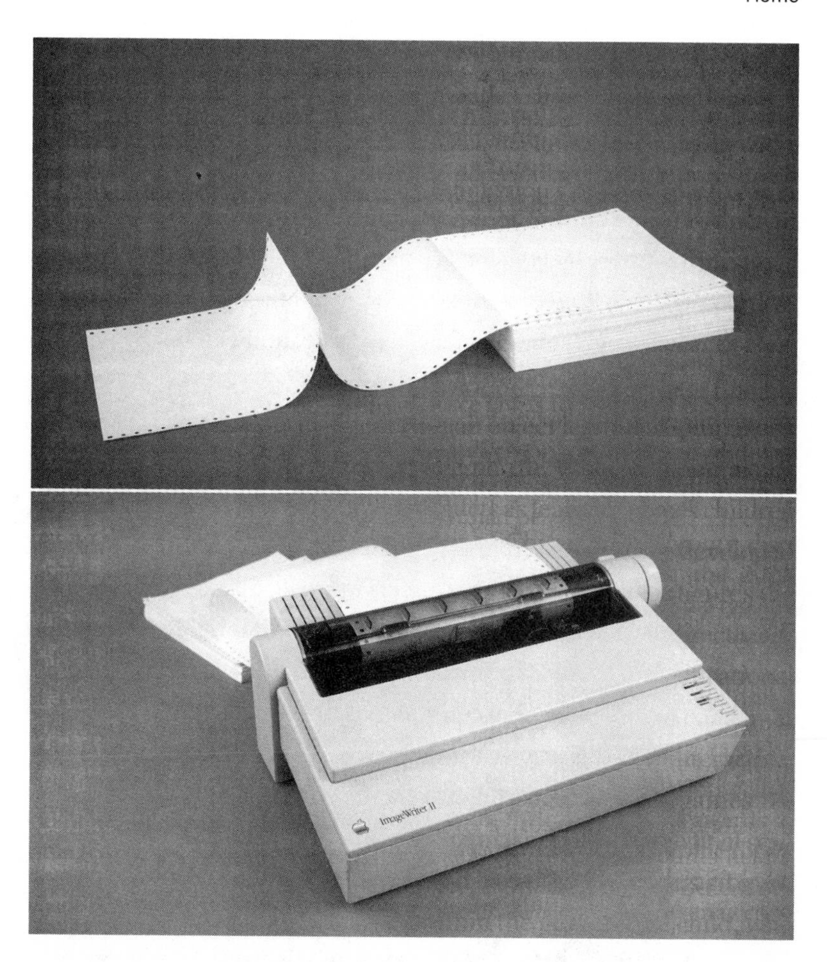

An example of computer printer paper from the 1980s (top), commonly used on dot matrix printers from the era, and that paper fed into an Apple ImageWriter printer (bottom). Paper for consumer printers was typically sold in stacks of continuous sheets with perforations that allowed individual pages to be torn off after printing. The holes on the edges of the sheets allowed the paper to be spooled through the printer; these strips were also intended to be torn off after printing. Images courtesy Tega Brain.

In Balsam's recollection, solving this problem was an "aha" moment in their product development, as it required a mental fluctuation between a two-dimensional digital image on screen and that image's expression as a three-dimensional object with multiple planes but a single surface. Using a graphics editing program on Kahn's Advantage to create print prototypes, Balsam eventually found his way to a solution:

> There's four quadrants on one side of the paper. If you fold the paper in four, you have one quadrant on the front of the card, another quadrant on the right inside panel, and there's also a quadrant on the back where you could put a little message. . . . I did the first version of that, a mockup on [Kahn's] graphics editor, turning things upside down by hand, using those two quadrants [and] folding it. I brought it to Marty, and I said, "Look, we can make this using just the front of the paper. We can make a greeting card that has a front and an inside."[61]

When printed on a sheet of paper, the greeting card page appeared "half upside down, half right side up in alternate corners," as *The Print Shop* reference manual stated ("Don't be alarmed," the manual also assured). The process of folding the greeting card, first top to bottom, then left to right, resolved these effects to create a 4.25 × 5.5 inch greeting card, as seen in the chapter's introduction.[62] Adorably diminutive, perfectly hand-sized, Balsam and Kahn's greeting card format gave human scale and physical weight to a computational object.

Important to this printing process was also the fact that Kahn designed the graphics, from the start, to be based on *printer* resolutions rather than *monitor* resolutions. As this book has illustrated, monitor resolutions of the period were quite minimal, constrained by the screen technology itself. Yet printers could lay ink on paper with a finer degree of detail than monitors could display, meaning images that might look chunky and blocky on screen could have some modicum of detail on paper. Thus *The Print Shop* wasn't printing what the user saw on screen; the screen image was only an approximation of the real image data held on the disk, a graphical shorthand to help users visualize their choices. In this sense, *The Print Shop*'s images, while clearly created by a computer, did not look rudimentary the way screen images did, especially when viewed at a distance.[63]

Memory doesn't serve to tell us how many other print formats Balsam and Kahn worked out before taking the revised prototype back to Brøderbund sometime in fall 1983, but there was enough there—enough images, enough text options, enough formats, enough graphical programming prowess, and, most important, enough potential for

An early prototype of *The Print Shop* greeting card layout, likely created on the NorthStar Advantage, ca. 1983. Courtesy Martin Kahn.

interaction—that Brøderbund moved quickly to support the development of Balsam and Kahn's new prototype.[64] In order to establish a traditional publisher-developer relationship, Brøderbund released Kahn from his position as a staff programmer, allowing him to form an equal partnership development studio with Balsam, which they called Pixellite Software.

When it came to choosing a platform for initial development, Balsam and Kahn didn't even consider anything other than the Apple II.

After all, it was the system on which Brøderbund had achieved its first real commercial successes in 1980. But the commitment to the platform was more than just publisher nostalgia. By 1983, the Apple II was the longest-lasting consumer-grade microcomputer model on the market. The system's long-term functionality was due not only to Wozniak's initial open-system hardware design but also to Apple's choice to keep investing in and improving on the platform—first with the release of the Apple II Plus in 1979 and then with the release of the Apple IIe in 1983. The Apple IIe ("e" for "enhanced") was a hardware revision of the original Apple II/II Plus system that lowered costs by reducing the number of parts while also making improvements to the system's keyboard, firmware, display, and other features.[65]

Yet the importance of the Apple IIe was not so much in what it did, technologically speaking, but in the longevity of the system's positioning relative to the broader microcomputer market. As Jerry Willis and Merl Miller advised in their 1984 buyer's guide:

> At a retail price of around $1,400, [the Apple IIe] offers no more hardware features than some machines selling for less than $600. We agree that the Apple IIe is overpriced when it sells for more than $1,000. You don't get much hardware for the price you pay. Yet when you buy an Apple IIe, you aren't really buying the latest computer technology. You are buying an opportunity—the opportunity to buy and use any of the thousands of programs that run on the Apple. More software is available for this computer than for any other machine in the world.[66]

With Apple keeping the same underlying open architecture as the original Apple II, new owners of an Apple IIe had immediate access to a massive supply of software and peripherals, most of which remained compatible across all platforms in the Apple II family. For software makers and publishers, this meant that the Apple II architecture remained a development tentpole for the consumer market, despite significant pricing and hardware competition from the IBM PC, the Atari 800, and the Commodore 64.

Moving forward to develop a prototype on the Apple IIe and buoyed by a small advance from Brøderbund along with the promise of a 20 percent royalty, the partners fell into loose roles: Kahn handled the programming and graphics; Balsam took on the more strategic concerns of overall program design (as well as serving as Pixellite's somewhat reluctant business lead).[67] The name, *The Print Shop*, came later, though it would be *owned* by Brøderbund, with Pixellite retaining only the rights to the software itself.[68] The product Brøderbund published nearly nine months later, in summer 1984, would achieve its remarkable success by placing a user, armed with nothing more than an Apple II and a printer, at the center of its world.

* * *

The talents and market strength of *The Print Shop*'s publisher, Brøderbund, went a long way toward ensuring effective publicity and retail positioning for the printer-based product, but it was the qualities of the software itself that would contribute to its enthusiastic uptake and meteoric success. Most obvious was simply what it did: it allowed users to create material objects. But perhaps more important was *how* it did that: through a carefully tuned user interface, which had been refined during the intervening months Balsam and Kahn spent in development. While the prototype of Perfect Occasion was built around the idea of superimposing text over graphics either produced by the user or created using the software's graphical tools, they determined that even that functionality was "too open ended."[69] Rather than focus on image generation or graphical rendering as the linchpin of their product revision, they shifted the creative emphasis to making choices among design elements, starting with choosing the print template, or "mode" (greeting card, letterhead, sign, or banner), before moving on to decorative features such as borders, fonts, font styles, graphics, and basic textual placement.[70]

To guide users through this broad assembly of choices, Balsam and Kahn organized *The Print Shop* as a series of dedicated, full-screen menus specifically intended to manage the user's sense of creative freedom.[71] Once the user chose a mode or template, the rest of the design experience was essentially on a rail. For example, a user designing a sign was required to make a choice, one design element at a time, in the following order: border, graphic, graphic size, graphic layout, and font, followed by a single screen for writing the message and setting text position, effects, and size. While users could go back to make changes in their selections, they could not choose where in the design process they started, nor could they alter the template itself. There was no way, for example, to use two different images on a banner, insert text on the inside cover of the greeting card, or rotate text on a sign. While some limitations were due to the hardware realities of making the software function on 48K, many of the constraints were intentional. As Balsam explains:

> [The Print Shop] doesn't ask too many questions of you, it doesn't demand anything. It works more by presenting you with a scenario that enables a part of you, an explorative or creative part, to move out into a little world. . . . It's not an open system, like a MacPaint. It's deliberately limited, but within those limits there is an infinite number of things you can do. It's a real delicate balance of deciding where the limits are, and where is the openness. I'd say that the hardest part of designing a program is deciding what *not* to do. . . . The concept of The Print Shop . . . is graphic arts for the non-artist and non-graphics oriented person. . . . You can't give our target audience that much of an open system—they'll get lost in it, and that would soon replace fun with frustration, totally defeating the purpose of the program.[72]

From the perspective of *The Print Shop*'s developers, the program's linear menus and various design constraints actually served to reduce the typical complexity of using a computer program. Balsam's observations thus reflect larger anxieties among the emerging computer user base as a whole. Microcomputers frequently frustrated their early, nonspecialist users, overwhelming them with technical language and complex demands that were believed to create a barrier between the user and the machine—the very kind of user alienation

Various menu screens from *The Print Shop*, running in MAME emulation on the Internet Archive (top). Right to left, top to bottom, these screens illustrate the step-by-step choices users made to create a sign in the program, ca. 1984. Screen captures courtesy the author.

that the marketing discourse of "home computing" was trying to shift in order to stimulate sales.

Software publishers and developers couldn't do anything about the fundamental challenges of using the computer itself. However, *The Print Shop* demonstrates that software concept and design meaningfully contributed to a growing sensibility of what it meant to adapt to at-home users who were neither professionals nor hobbyists. *The Print Shop*'s user experience design thus served to productively reposition where the experience of creativity happened for the user. Instead of being confronted with, and overwhelmed by, a system that might be programmed to do anything—which was the source of creative excitement for hobbyists—*The Print Shop*'s users produced creative experiences by combining a limited number of print formats and design elements in various permutations, all of which could be printed out and carried into the user's life.

The accessibility of both the program and its documentation was frequently highlighted in reviews of *The Print Shop*. "This is not a game," Margot Comstock Tommervik wrote in her June 1984 review of the program in *Softalk*. "Repeat, this is *not* a game. It shouldn't have everyone in the family lining up to play with it, it shouldn't draw crowds at gatherings, and it shouldn't be incredibly addicting. But it does, and it is."[73] For Tommervik, *The Print Shop*'s "self-explanatory menus," in which "all choices are graphically represented on the screen as you make your selection," were an example of the software's "rampant" "consideration for the user."[74] Similarly, a March 1985 review in *COMPUTE!* emphasized that "it's not an exaggerated claim" that users didn't need to read the manual: "The program is pretty thoroughly error-proof."[75]

In a world of backup disks and thumb-thick manuals, reference cards, and changing disk operating systems, where many new users were discovering that programming was hard, and often not that fun, and all the novel things you might do with the computer were far more challenging than you realized, the appeal of *The Print Shop* was that it managed to make computers do something complex and creative in a way that felt so dead simple, you didn't even need to read the instructions. Retailing for only $49.95, the program was also an affordable bright spot in what was otherwise shaping up to be a grim year, as the consumer computing bubble began popping industry wide.

But it was also the program's *output*—its capacity to let users create objects with material existence—that proved such a powerful point of attraction for *The Print Shop*. For so much of the "home" software market, computerization itself was the novelty, full stop. But by not treating the computerization of the greeting card as an end state, Brøderbund's feedback pushed Balsam and Kahn to imagine the computer as simply an intermediary between input and output—between having an idea and allowing it to exist in the world.

While the private nature of home computing makes it hard to know what exactly people did with *The Print Shop*, a fair bit of evidence exists in the product's documentation, as well as the enthusiastic coverage the program received. A 1985 feature story on Pixellite Software reported that *The Print Shop*'s fan mail folder was over an inch and a half thick, documenting uses that ranged from teaching spelling to assisting with cognitive and art therapy (though the details are not described).[76] Images in *The Print Shop*'s reference manual offered examples clearly intended to inspire users to think of what in their lives could benefit from *The Print Shop*'s flair. The manual features cards for anniversaries, holidays, and thank-you notes; posters advertising garage sales, senior proms, piano lessons; printed to-do lists and "no smoking" signs; small business letterhead; and banners honoring birthdays, weddings, and graduations. And while *The Print Shop*'s graphical capabilities were no match for professional printing services, Balsam claimed that a few users had managed to set up modest side hustles using the program—one selling banners to local organizations and another claiming to make "monogrammed napkins for restaurants."[77] By 1987, Doug Carlston reported that Brøderbund had received "thousands of letters" documenting unanticipated uses, including "a stencil for painting the name on a sailboat, . . . gift wrapping, Christmas ornaments, party hats and to frame pictures."[78] Users' efforts to stretch the software toward these unanticipated purposes reflect what was at the heart of so many marketing efforts for home computing: getting people to imagine how a computer could intervene in their lives, even when what they were doing with the computer could be completed just as simply by other means.

While scattered and incomplete, computer software sales charts also provide a compelling impression of *The Print Shop*'s popularity among consumers. *Softalk* folded before it could document much of *The Print Shop*'s success within the Apple II market, but the entertainment industry trade magazine *Billboard*, which began tracking computer software sales in October 1983, tells the story well enough.[79] *The Print Shop* first appeared on *Billboard*'s Home Management Top 10 chart in the July 28, 1984, issue, debuting at number four and rising to second place the following week; it was the only software in *Billboard*'s "home" category that wasn't a word processor, a home finance package, or an information management suite. Notably, *Billboard* tracked sales across a variety of microcomputing systems, including the Apple II, the Atari 400/800, the Commodore 64, and the IBM PC—making *The Print Shop*'s immediate status as a top five (and soon-to-be number one) best seller even more remarkable given that it was initially available exclusively for the Apple II and was competing against software ported to multiple platforms. Porting *The Print Shop* to the Atari 400/800 and the Commodore 64 took nearly a year (reprogramming the software for Commodore's most popular printer was apparently quite a challenge), but the program's popularity within the home software category never swayed.[80] *The Print Shop* remained on *Billboard*'s Home Management Top 10 for 113 weeks—and would almost certainly have lasted longer had *Billboard* not discontinued its computer software charts in fall 1986. But in the nearly three years that *Billboard* published its sales charts, *The Print Shop* was beaten for longevity by only a handful of other products, such as Brøderbund's *Bank Street Writer*, Davidson and Associates' *Math Blaster*, and Microsoft's *Flight Simulator*. No other piece of home software came close.

It was for these reasons that Doug Carlston, from atop his perch at Brøderbund, would come to identify *The Print Shop* as "evergreen," a quality he had already seen in *Bank Street Writer*. "These were products you could sell year after year," Carlston recollected. "We were just beginning to think in those terms."[81] In a market ecology where best sellers typically rose and fell in a matter of months, or even weeks, an evergreen product could transform a company, ensuring ongoing revenue while it developed other products. Furthermore, *The Print Shop*'s design lent itself to selling additional graphics packages featuring more borders, fonts, backgrounds, and pictures—making *The Print Shop* the rare example of a 1980s software package that successfully spurred sales of a complementary good.

In this sense, *The Print Shop* wasn't just a product; it was a brand. In 1987, Brøderbund announced that average sales of this single product represented an astounding 4 percent of the entire consumer software market for the year.[82] By 1988, *The Print Shop* sold its millionth copy.[83] It may not have been the "killer app" for the home computing market in the way that *VisiCalc* was for the business market, but it didn't need to be. The power of *The Print Shop* lay not in giving people a reason to own a computer—better reasons existed—but in enhancing the sense of intimacy people had *with* their computers. Computer experts and novices alike took the program's straightforward approach to print-based creation and ran with it. As a result, programming and hardware hacking no longer occupied the sole horizon for computing hobbyism: printing your own greeting cards, fashioning wrapping paper, and making birthday banners could become hobbies in themselves. This shift reflected the broader turn exemplified in the notion of "home" computing—that the computer could serve a world of potential needs as unique as its users and thus could find a place in people's everyday lives.

* * *

In the 1987 introduction to *The Official Print Shop Handbook*, a three-hundred-page compendium of program-specific designs, customized graphics, and hacks and tricks, its coauthor, Randi Benton, recounted the moments when it became clear to her that *The Print Shop* was making a dent in the world. The first was when she was on a vacation in a small village in the South of France, a hilltown of cobblestones, iron-wrought signage, and "narrow, winding lanes."[84] Yet in a tiny old stone museum, she found flyers for an art exhibition that had been made with *The Print Shop*. Shortly after returning from this trip, *The Print Shop* (Benton believed) made another appearance, this time on the national stage.

> It was early October, better known as baseball playoff season in our family. During Game 4 of the American League playoffs the cameras zeroed in on an oversized sign in the stands that read:
>
> ROSES ARE RED
> VIOLETS ARE BLUE
> THE ANGELS IN FIVE
> I GOT SERIES TICKETS TOO!
>
> I spotted it right away. The sign was created from four Print Shop banners stacked one on top of the other.[85]

Screen capture from Game 4 of the 1986 American League Championship Series between the Boston Red Sox and the Los Angeles Angels. Fans in the stands display a printed sign created with *The Print Shop* competitor, *PrintMaster*. Screen capture by author. Video posted on YouTube by user Classic MLB1, February 22, 2019: https://www.youtube.com/watch?v=zjXKixxvoFI

For Benton, these incongruous moments of finding *The Print Shop* out and about in the real world were a testament not just to the software's utility but also to the world of clever ideas users brought to its implementation.

But Benton made an error—though an understandable one given that the image flashed on the screen for no more than a few seconds. There was a homemade printed banner in that game, just as she described, but it was produced with the Unison product *PrintMaster*, a *Print Shop* clone so similar in look and feel that Brøderbund successfully sued for copyright infringement in 1986.[86]

The court case would create a precedent for extending copyright protection to user interfaces, but for consumers, the differences between *PrintMaster* and *The Print Shop* were six of one, half dozen of the other. What *The Print Shop* had created was no longer just a product, but a kind of sociocultural meme. If *VisiCalc* engendered a "spreadsheet way of knowledge," *The Print Shop* might be regarded as having catalyzed a way of seeing our lives as available for creative and expressive augmentation through the program's visual, programmatic, and material logics (a computer way of making? a printerly mode of creativity?). In Benton's error of identification—an example so innocuous, yet also so fleeting it could only be vetted nearly forty years later thanks to the unquantifiable mass of content supplied by YouTube's users—we see the collective effect of programs like *The Print Shop* as a vernacular digital practice, a way of thinking and doing computing that had nothing to do with code. What could possibly be more personal than *The Print Shop*'s way of "doing" creativity becoming situated, repeated, and templated within us, and one that assumed we knew nothing about computing at all?

And somehow it never ended. After *The Print Shop Companion* (1985), *The New Print Shop* (1988), *The Print Shop Deluxe* (1993), *The Print Shop Deluxe Companion* (1994), and other installments, the version ticker eventually reset to one, and *The Print Shop* is still being sold as of spring 2022, deluxe version 6.0, available for download from present-day Broderbund's website for the uninflated price of $49.99 (the company name lost its stroked "ø" at the turn of the twenty-first century).[87] Pixellite Software had been separated from the project since the early 1990s, after frustrated dealings with Brøderbund. Yet while Balsam and Kahn owned the program, Brøderbund owned the name. Pixellite was free to exit its publisher arrangement with Brøderbund, but the brand Balsam and Kahn had helped build would not come with them.

Trudging onward, Balsam and Kahn took their code base and institutional knowledge to other publishers, pushing the program into its own series of reincarnations.[88] In 1992, it reappeared as *Instant Artist*, published by Autodesk in a quickly aborted attempt to enter the consumer software market. From there, it ambled on to Maxis, the famed developer of *SimCity*, which published it as *Print Artist* in 1994, branding it as a "personal creativity" product that came with "the artist built in."[89] In 1995, Sierra On-Line—then a behemoth of the consumer software market—stepped in to acquire Pixellite and its star product in full and leaving the program's developers with a healthy buyout. Under Sierra's "home" division, it would become the enormously popular *Sierra Print Artist*, a program also still somehow in circulation, available for purchase from Nova Development.

The ecosystem that Apple cultivated between software developers and users, which allowed developers to forge new imaginations for computing by lowering the difficulty threshold for software use, turned out in the end to have more staying power in defining what personal computing would look like than the out-of-touch imaginings of hard-core hobbyists or the prescriptive visions of retailers and advertisers and industry prognosticators. In many ways, both imagined too much for their users. Neither the computer revolution nor computational domestic bliss did much to move units; what most people wanted was not to have to think about what was going on inside their Apple. The tremendous value of understanding *The Print Shop* as part of the historical record is in recognizing that it performed none of the allegedly valuable functions home computing was supposed to be good for, yet somehow came to define what was most exciting, creative, and playful about the entire software genre. In doing so, it both dismantles any assumptions we might carry about the obviousness or inevitability of home computing and affirms that home computing held fascinating potential. A world of users who didn't know a printer driver from an interface card was not what most hobbyists hoped for when they dreamed of the computer revolution. For better or worse, this was computing as consumption, in which the hobby no longer revolved around the computer itself but around mastering its myriad applications.

7

Education

Snooper Troops

The April 1985 issue of *Creative Computing* dedicated itself to a not at all straightforward question: "Educational Computing: Where Are We Now?" *Creative Computing* was an appropriate venue for such ruminations. Founded in 1974, before there was a microcomputer revolution to speak of, the magazine had long been a resource for educators who wanted to use computational systems for teaching.[1] While the magazine had shape-shifted over the subsequent decade to reach an audience beyond teachers, it retained a special emphasis on the educational applications of computing—carrying the torch, long before other periodicals, for the capacity of computers to revolutionize learning.

So where *was* educational computing around spring 1985? *Creative Computing* never gave a straightforward answer, but we might intuit one from the issue's main feature, a "software directory" titled, "Goodbye, Little Red Schoolhouse." The article's feature image is *Little House on the Prairie* meets the Age of Information: a one-room schoolhouse built from software, with floppy disks for shingles and walls tiled together from software packaging. Children are nowhere to be found in the structure; the school's only occupant is a smiling Macintosh, Apple's first consumer-grade desktop microcomputer with a graphical user interface.[2]

As for the directory itself, it's hard to imagine what the reader of 1985 would have taken away from this ten-page feature. Companies are listed in alphabetical order, each accompanied by a modest summary of its most popular products, but no selection criteria are provided for what "educational" might mean. SAT prep programs shoulder alongside sixth-grade math equation drills, software introducing colors and shapes to preschoolers is listed next to adult tutorials on how to use databases. Certainly, there was a lot of software. But what did it do? Whose needs did it meet? How could you locate what *you* needed? And was any of it worth buying? In the editorial preceding the directory, the education technology beat reporter Betsy Staples acknowledged that parents and educators were adrift in an "uncharted sea of highly touted, expensive, and potentially useless software."[3] Her primary advice was ironic: parents and teachers needed to teach themselves how to make sense of it all.

The sheer scale of unguided overabundance reflected in "Goodbye, Little Red Schoolhouse" illustrates the perplexing state the microcomputer education software market had achieved just a few years into its existence. Experimentation with computer-supported learning can be traced to the 1960s, but the perceived necessity of computing in schools took on new forms in the early 1980s, as the dropping costs of hardware and the expanding market for software intersected with new anxieties about whether America was ready for the demands of the impending Information Age. As wages stagnated, inflation roared, and unemployment clipped ever upward, techno-futurists, academic researchers, and computer hustlers alike claimed that the only way to deal with the changes wrought by technology was to allow ourselves to be remade by even more technology. American educators, parents, administrators, and policy makers thus internalized a catalytic sense of urgency on the topic of computer literacy—an urgency no doubt amplified by the impression that there was a ready solution, already available, in the low-cost, desktop microcomputer. Over half a million microcomputers entered the nation's public schools between 1980 and 1984, shifting the proportion of schools "using at least one microcomputer for instructional purposes" from 18.2 to 85.1 percent in just four years.[4] As *Popular Computing* put it in 1983, "Schools are in the grip of a computer mania."[5]

If this surge left teachers overwhelmed, parents frustrated, and administrators desperate to find new elasticity in old budgets, the microcomputer hardware and software industries saw nothing but opportunity. Hardware manufacturers tripped over themselves in the race to install their computational brand recognition before a generation of American children. By 1985, Apple Computer would dominate just over half the US education market—more than double its nearest competitor, RadioShack, and three times more than Commodore.[6] Software, likewise, became a bonanza, as long-standing textbook publishers moved to add software to their listings of instructional materials, and, simultaneously, new companies targeting that same market blossomed by the dozens.

And in the midst of all this, an at-home market for educational software found fertile ground for growth. Unfettered by the demands of curriculum-specific benchmarks, consumer-facing publishers appealed directly to well-off parents anxious to supplement whatever limited exposure their children had to computers in school. One of those home education software publishers was Spinnaker Software, founded in 1982. Two of its first products, a pair of deductive reasoning mystery games sharing the name *Snooper Troops*, were released that same year. *Snooper Troops Case #1* and *Case #2*, both designed by independent software developer and middle-school teacher Tom Snyder, would become some of Spinnaker's most popular and best-remembered products. The relationship between *Snooper Troops* and Spinnaker embodies much of the urgency, and financial interest, underlying the abrupt boom of the educational software market while also illuminating the distinctions in how software designed for curricular, school-based instruction and at-home use were developed.

Founded in 1982 by Bill Bowman and C. David Seuss, two former Boston-based management consultants, Spinnaker approached the educational software market with the ethic of a confidence game. Unlike companies such as Sierra On-Line or Brøderbund, this wasn't a business bootstrapped from homebrew enthusiasm or driven by programmer expertise. Spinnaker was conceived as a business plan and venture backed to the tune of $800,000 before it ever had a product. From Bowman and Seuss's perspective, the retail software market's real impediment to growth was that it lacked the "rigors of consumer marketing"—a problem they believed they were uniquely poised to solve, regardless of whether either man had ever typed out a line of BASIC.[7] Their corporate philosophy was simple: build a consumer pipeline and outspend every competitor on marketing.

Yet while Spinnaker's marketing strategies undoubtedly raised the profile of *Snooper Troops*, the game itself was more than the sum of its advertising. It was originally conceived to be played by groups of children across multiple computers under the instruction of a single teacher in a classroom session, but Snyder had to redesign the game as a single-player experience when he was approached by Bowman and Seuss to make a product for the home market. Yet much of what made the game notable—its interactive and exploratory qualities, the way it handled the methodical parceling out of information, the way it relied on attention and notetaking—were techniques Snyder had drawn from earlier work developing curricular educational software, as well as his own teaching philosophy about the importance of facilitating consensus among children. In this sense, *Snooper Troops* is best understood as an unruly offshoot of a *development* philosophy that never intended for a single child to be sitting in front of a single computer—which was, uncomfortably enough, the very *consumer* philosophy Spinnaker saw as critical to its success. While the respective ethos of Snyder and Spinnaker may have been at odds, Spinnaker and *Snooper Troops* evidence the broad spectrum of stakeholders and ideologies that bore up the American educational software market.

* * *

In January 1978, the twenty-eight-year-old Tom Snyder brought home an unexpected addition to the Cambridge, Massachusetts, apartment he shared with his girlfriend, Anne Waddington: a TRS-80 Model I.[8] He described the purchase as a bit of "retail therapy" immediately following a botched pitch to Parker Brothers on the North Shore of Boston. He had been there trying to sell the game acquisitions department on an interactive device he had built for his classroom but was told on arrival that he missed his appointment by a day and there was no hope of rescheduling for months. Dejected on his drive back to Cambridge—"just not believing what personal, psychological self-defeating mechanism had orchestrated this epic failure"—he pulled over at a RadioShack, intending merely to nurse his wounds by browsing electronics.[9] Instead, he emptied his savings and walked out with what was, at the moment, the best-selling microcomputer in the country.

Snyder had never seen a microcomputer before and knew nothing of 1970s computer hobbyism or hacker culture, but he wasn't totally out of his depth. Like so many of the men who later found themselves in the hardware and software industries, he had been an electronics buff in his boyhood. Snyder holds crystalline memories of finding Claude Shannon's master's thesis on Boolean logic in his Dedham, Massachusetts, prep school library, and he even built a homebrew, relay-based computer in his parents' basement in the early 1960s.[10] In high school, he phased out of the obsession, trading computers and electronics for rock 'n' roll and girls ("I erroneously thought I looked a little bit like Paul McCartney," he gently recalls), but all of that electronics knowledge was still with him, humming under the surface. As he explained to Waddington when he got home, he was certain he could do *something* with the machine, even if he wasn't sure yet *what*.

Educational software designer Tom Snyder, working with children playing games on an Apple II, ca. mid-1980s. Courtesy Tom Snyder.

"It was like I said, 'Honey, I sold the cow, but I bought these magic beans.'"[11]

His plan, insofar as he had one, was to use the TRS-80 to extend his pedagogic—and entrepreneurial—goals. Snyder was a middle-school science teacher at Shady Hill School in Cambridge, a job he had come to rather circuitously after obtaining a French literature degree from Swarthmore College, cutting a record with Capitol, and playing gigs in the early 1970s in Vermont ski country. Growing tired of the hustle, Snyder enrolled in an apprenticeship graduate program at Lesley College in 1974 and took his permanent position at Shady Hill in 1975, where he met Waddington a couple of years later. Snyder's teaching philosophies would be heavily influenced by Waddington, or more accurately, by the spiritual heritage they shared: both had ancestral ties to the Religious Society of Friends, also known as the Quakers, a Protestant Christian sect known for its emphasis on community, pacifism, and individual communion with God. While Snyder had been familiar with Quaker tradition through his upbringing and his time at Swarthmore (a Quaker-founded school), he took to the tradition with renewed interest after striking up his relationship with Waddington.

During this time, Snyder was especially drawn to the Quaker model of consensus-based decision making, in which the group's achievement of unity through communal discernment produces understanding "superior to the previously held opinion or judgement of any single member."[12] While Snyder didn't aim to convey an explicitly religious philosophy to his students, he found the consensus model a powerful way to engender communication and discovery among students and spent many of his after-hours developing activities to facilitate such engagement.

To make the consensus model work as a curricular technique, Snyder first had to develop scenarios in which predetermined information could be divvied up among multiple students so that group discussion would enable them to produce a correct observation. His first experiment with this style of teaching involved no computer but foreshadows the technical curiosity that made Snyder especially open to exploring educational computing: in 1977, he began prototyping an electromechanical relay device intended to demonstrate how the five senses work together. Dubbed the "Personk," the device was a five-armed wooden contraption, spiderlike, with headphone jacks for output and dials for input, allowing each child to privately receive a unique piece of sensory information.[13] Only by sharing that information with each other could the group of children determine what they collectively "sensed" and thus make decisions about how to move through the world.

While these explorations may have served to educate his students, Snyder also recognized and pursued their commercial potential. Personk was the very device he had arranged and then failed to pitch to Parker Brothers in 1978—the same failed pitch that led him to RadioShack, to the TRS-80 and everything he might do with it. Here Snyder's technical expertise, unique pedagogic style, and personal creative and entrepreneurial energies became mutually reinforcing. Contrary to the way popular accounts of this era often narrativize the effects of obtaining a computer in the 1970s or 1980s, seeing the TRS-80 did not "inspire" Snyder to pursue educational computing, as if the microcomputer could manifest desires where none had existed before. Rather, the capacities of the TRS-80, which Snyder understood given his childhood background in electronics, were a comprehensible extension of activities he was already exploring in the classroom. The benefit of the microcomputer was that it could evaporate the time and expense of building customized devices, allow for quicker iteration, and command more information than a relay-based device reasonably could. This context helps us understand how Snyder rationalized buying the microcomputer: far from seeing it as just a technological toy for his own amusement, he would almost certainly have been imagining some kind of commercial application.

Snyder dabbled with a prototype of Personk for the TRS-80 but wound up sinking his energy into a more ambitious project, which started as a pen-and-paper navigation simulation but was quickly computerized as Snyder taught himself BASIC from the TRS-80 handbook. The simulation was designed for his geography class and was scaffolded on the colonial logics of the Age of Discovery: students had to navigate their ship across the ocean to find a New World and then return home with a map and riches. As the children traversed the ocean in groups of four or five, they scratched away at simplified versions of the math that made long-range colonization possible, continually discerning their location at sea based on environmental and astronomical data: one student was in charge of ocean depth, another had to know the position of the stars, another monitored the trade winds, and so on. Together, they plotted a course that none could find alone. As they updated their bearings, each group of students came to Snyder for new input. "I would roll the dice," Snyder recalls, "and had a little calculator and I would quickly jot down some information, fold it up on a piece of paper, and send them back to their desk."[14]

Snyder found such techniques effective but "teacher-intensive," requiring extensive preparation as well as on-the-fly command of all the variables constituting the game state. The benefit of Personk was that the relays inside the contraption handled the generation of output and implementation of input, but Snyder's navigation simulation was too complex to efficiently or cost-effectively construct in relays. Snyder didn't need a lot of computing power, but he did need more than could be reasonably fabricated from the more physical computing components he had experimented with previously. The TRS-80 was an in-box solution to exactly these conundrums.

Snyder began computerizing his simulation later in 1978, writing BASIC code to maintain and update game state, handle random number generation, and generally make the experience more immediate and enlivening. Using the computer also allowed him to more easily prepare similar simulations for other social science topics; after the initial geography program, Snyder produced a repertoire of five social science simulations on the topics of energy use, archaeology, geology, and civics. Importantly, Snyder did not conceive the computer as a way to replace teacher-led instruction. Rather, the amount of calculation, analysis, and record keeping his simulations required of their participants further entrenched him as central to the learning experience. The computer was a machine for handling simulation math, but decisions had to be made by the children themselves, and Snyder's presence was essential to facilitating their dialogue and helping them through rough spots.

If there were any concerns about a young teacher bringing a computer into the classroom, Snyder doesn't recall them—a comfort that reflects the urban, well-educated social milieu of the children he taught at Shady Hill. In Snyder's recollection, his students were mostly the children of "Harvard and MIT faculty . . . or old Cambridge money," precisely the kinds of parents either wealthy enough or educated enough to have some recognition of the computer as a tool of the future. And Snyder was a competent showman of his own educational efforts, demonstrating his creations at parents' nights and helping them get excited, as their children were, about this intersection of play, technology, and learning.

Mixed into the student body were also a few of the kids from Boston's investor class—a critical context for Snyder's future prospects. One night in 1978, Snyder was shopping at a liquor store in Harvard Square when he ran into a man named Jere Dykema, a lawyer turned investment manager, whose son was in one of Snyder's classes.[15] Dykema recognized Snyder and struck up a conversation, eventually accompanying Snyder to dinner. Over beers, Dykema listened to all of Snyder's entrepreneurial tales—about Personk, the Parker Brothers screwup, the TRS-80, his simulation games, even his not-quite-done music career. We'll never know what precisely Dykema saw in Snyder, but he must have recognized a furious energy there, a willingness to create and work and make to the point of exhaustion. And this was sometime in 1978: elsewhere in Cambridge, Dan Fylstra was shipping orders of *Microchess* from the living room of his apartment; Dan Bricklin and Bob Frankston were perhaps figuring out how to pivot their spreadsheet program for an Apple II; and on the other side of the country, the Apple investor Mike Markkula was turning little computers into massive money. Whether Dykema knew some of this, or all of it, or none of it, he saw something in Snyder, the kind of insatiable energy that brute-force capitalism demands. At the end of their dinner, he made Snyder an offer, or so Snyder recalls: "If I gave you $30,000 right now, can I have 30 percent of your business?" Not any specific business or product, not Personk or the music or whatever Snyder was up to with his little simulations. What Dykema wanted to buy was *Snyder*.

The offer was close to five years of Snyder's salary as a teacher. Lacking a preexisting bankroll or family financial support (or any other men offering him tens of thousands of dollars over beers), Snyder readily accepted. Though he maintained a part-time position at Shady Hill, Snyder established a company name for himself, Computer Learning Connection, and began regularly meeting with Dykema to strategize on his projects. Dykema closed the gap between Snyder's desired grasp and his actual reach—pulling the bough down just enough for Snyder to grab the fruit, knowing, of course, that three of every ten apples would be his.

* * *

Leveraging his substantial contacts across the business world, Dykema brought in a consultant to suggest publishing options for Snyder's simulations and was directed to pitch to the international publishing company McGraw-Hill. While primarily known today as an educational publisher, McGraw-Hill was, in the late 1970s, a massive information services conglomerate with business interests spread across publishing (including books, periodicals, and audio-visual media), financial services, and broadcasting operations. The company excelled at managing flows of information, knowing where to exact a toll, and evaluating where future flows (and their accompanying tolls) might be forthcoming.[16]

Snyder and Dykema approached McGraw-Hill sometime between 1978 and 1979—roughly the same time McGraw-Hill acquired *BYTE*, the tentpole periodical of hobbyist computing. The acquisition of *BYTE* was more than simply the purchase of a popular computing magazine; it was one part of a multifaceted effort to establish consumer tollbooths around what McGraw-Hill predicted would be the technology of the Information Age. By 1979, McGraw-Hill had several such tollbooths under way: in addition to *BYTE*, the company had launched *onComputing*, targeted at beginner at-home computer users; acquired a small-scale publisher of computing books and manuals; and was developing a portfolio management subscription service for the TRS-80.[17] Unlike McGraw-Hill's competitors, which largely entered the industry four or five years later, expecting to ride a software gold rush that never came, McGraw-Hill didn't need overnight results. It was enough for them to be steady, to wait out the turmoil, like an ant packing away its grain for the winter. The company would take a similar approach to its entry into microcomputing, investing early and drawing on synergies across multiple departments to enter a realm of information services previously untapped: educational computing.

While educational computing can be traced to the 1960s—most notably with the Dartmouth Time-Sharing System (DTSS) and the Minnesota Educational Computing Consortium (MECC), both extensively documented in Joy Lisi Rankin's *A People's History of Computing*—these efforts were typically built around time-sharing networks and were by no means ubiquitous on a national scale.[18] As Victoria Cain notes in her history of educational media, *Schools and Screens: A Watchful History*, "accidents of geography and demography determine[d] computer access in K-12 schools" during this time, as children closest to university towns or high-technology hubs benefited from disproportionate access to time-sharing systems or other forms of computing exposure (a theme that recurs in this book as well, from Wozniak to Bricklin to Balsam to Snyder himself).[19] For example, DTSS grew from the research pursuits of professors at the Ivy League Dartmouth College, while the public will for MECC and earlier regional explorations of educational time-sharing benefited tremendously from the fact that the Minneapolis–St. Paul region was a hub for numerous computing businesses, including 3M, Univac, Honeywell, and Control Data.[20] While publishers like McGraw-Hill had trafficked for years in non-textbook media offerings such as films and slides, the highly localized and regionally idiosyncratic conditions that shaped the integration of time-sharing systems into districts and individual schools would not have made for a particularly appealing educational services publishing market.

The microcomputer, however, shifted this economic and industrial calculus by making access to computer power less expensive for individual schools, at least on the surface. Time-sharing was cost-effective only when expenses could be distributed across many users, which included buying or leasing minicomputers or mainframes, buying terminals, and paying connection fees, as well as network maintenance and employee costs. Microcomputers, at least at first glance, seemed comparatively cheap and could be purchased without the oversight of the state or regional governing body. Much in the same way Apple IIs made their way into Wall Street offices by dodging the governance of data processing departments (as discussed in chapter 3), microcomputers could (and did) slip into schools through the front doors, as Snyder's own tactics demonstrate. Numerous articles from the early to mid-1980s document parent-teacher associations hosting raffles and bake sales in order to raise money to buy microcomputers, while in 1982, the *New York Times* reported on a twenty-six-year-old California schoolteacher who took on $15,000 in personal debt to buy computing systems for his students.[21]

While such moves no doubt felt invigorating to the teachers making them, they express, as a whole, a neoliberal shift in the approach to funding computing in the classroom. The perceived cheapness of microcomputers mattered in an environment where federal funding for educational computing initiatives was waning, even as computing technology was increasingly seen as a central part of American innovation and key to maintaining the nation's standing in the global economy.[22] In this context, microcomputers appeared to be a low-cost, quick-fix solution to the problem of having to do more with less. What school district wanted to coordinate a time-sharing network if it believed it could get similar results for just a few thousand dollars, on its own terms and on its own timeline? And while the turn toward microcomputing left many teachers to educate themselves about computing and its place in the curriculum (rather than being able to rely on the institutional knowledge of a larger time-sharing network), this seemed to do little to tarnish microcomputing's promise. Microcomputing entrepreneurs were happy enablers of the technology's supposedly inherent ability to inspire students to achieve individual success, and many teachers were eager to do this (routinely unpaid) work.[23] As Cain writes, "Personal computers intrigued many of the same factions that had championed educational television in the 1950s . . . teachers enamored by new technology, administrators hoping to improve learning efficiency and reduce teaching costs, policymakers who saw technology as the key to modernizing US schools."[24] Educational microcomputing was thus a fraught bargain compared to its precursor in educational time-sharing: it offered a value proposition that could only be a value proposition in a country wanting to improve academic results while reducing and privatizing academic funding. The outcome was a kind of nationwide educational paradox: the only way for the United States to succeed as a collective nation on a global stage was for schools to internalize an individual sense of responsibility for such results.

Yet the shift in education from a handful of distributed, collective time-sharing networks to a potentially vast sea of independent desktop computing systems was precisely what optimized the economic conditions for a company like McGraw-Hill to begin publishing educational software targeted at the school market. Snyder and Dykema went together to make the pitch to McGraw-Hill, Snyder hauling his TRS-80 all the way to New York City and up a skyscraper. While there were certainly plenty of schoolteachers plugging away on projects for their classrooms, none were showing up with a suite of finished, play-tested software and the backing of an experienced financial executive. So in a rented suit, while fending off an anxiety attack, Snyder sold the hell out of his software to "a roomful of guys with wingtips and pictures of their families on their desks."[25] However unfamiliar that classically Midtown corporate culture may have seemed to Snyder, what he showed them resonated with their executive self-interest: here was another tollbooth, another way to craft a monetary on-ramp by strategically controlling information within their incumbent distribution networks. In other words, whether it was textbooks or slides or film reels or software, McGraw-Hill knew how to install its products in school systems.

Snyder's software simulations would become known as *Search Series*, a five-program suite targeted at middle-school social science teachers.[26] For Snyder, it was the beginning of a career as an educational software developer, a chance at making real money, a chance at having a real business. For Dykema, it was a chance for Snyder to produce a return on investment. For McGraw-Hill, it was one of the earliest entries in the company's new conception of "courseware"—a term first appearing in its 1981 annual report to shareholders, a report tailored to affirm how McGraw-Hill was keeping its pulse on the "communications revolution."[27] In this sense, Snyder, Dykema, and McGraw-Hill were all responding to the same subtle forces in the world, creating a self-reinforcing feedback system between invention and finance.

Aside from producing all the documentation for *Search Series*, Snyder had to meet one further requirement to secure his contract with McGraw-Hill: he had to port the software from the TRS-80 to the Apple II. While the late 1970s were still early days for microcomputers in the classroom, the benefits of the educational market would have been obvious to hardware manufacturers: bulk sales with trickle-down effects, as parents might be influenced to purchase the same microcomputer for their home that their child used at school.[28] Apple was aggressive about these prospects early on, winning a bid in 1978 to become the primary provider of microcomputers for MECC—the same consortium that had made Minnesota a leader in educational time-sharing.[29] The deal had significant knock-on effects. As Rankin notes, "When MECC adopted Apple IIs, the consortium gained a new role, that of software translator from time-sharing BASIC to Apple BASIC."[30] In other words, MECC, which had a massive catalog of software generated by the educational time-sharing user community, began converting its creations for the Apple II, creating a bank of educational programs on the system that would wind up spurring Apple II purchases elsewhere. Schools and districts were understandably reluctant to invest in microcomputer purchases if there was not a preexisting supply of ready-to-use software. Though it's unclear to what extent McGraw-Hill was aware of these dealings, it was informed enough to believe that it was necessary to release *Search Series* for the Apple II. The incumbency of the Apple II in the educational software market was thus not purely a consequence of the microcomputer itself but also of the Apple II's strategic positioning within the broader hardware and software ecosystem.

* * *

It was December 1, 1981, and Bill Bowman didn't know who he was having lunch with. He had arrived that day at TA Associates, one of the largest independent venture capital firms in the nation, with plans to meet with the firm's manager—only to be told he wasn't available. Instead, someone named Jacqui Morby would be meeting with him.[31]

Bowman was apprehensive. A business executive and former marketing consultant in his early thirties, Bowman was at TA on a fishing expedition to get feedback on a loose plan he had to cofound a minicomputing software startup. Bowman and his partner, C. David Seuss, had the kind of pedigree we should no longer find surprising at this point in our tale: both were graduates of Harvard Business School in the 1970s (Bowman was in the same cohort year as Personal Software founder, Dan Fylstra) who met while working at the prestigious Boston Consulting Group, doing market analysis and strategy for what Bowman termed "pretty boring, 2 or 3-percent-growth industries."[32] Bowman didn't actually think he and Seuss needed funding; he had gone to TA out of obligation to a friend who had brokered the introduction. This made him all the more uncertain about having his meeting pushed off onto someone else, especially because, as he would later admit, "I didn't know who she was—a secretary or what. I'd never heard of her."[33]

As it turns out, Jacqui Morby was not a secretary. She was one of TA's vice presidents, and the force almost single-handedly leading the firm into funding the computer revolution.[34] In the late 1970s, she had cut her teeth finding undercapitalized prospects in the mainframe software industry, but more recently, she had helped nail down a four-way venture capital firm deal to invest in Gary Kildall's startup, Digital Research, maker of the popular CP/M operating system.[35] To understand Morby through the deals she made would be to characterize her as industrious and tenacious, but she was also being pushed along by larger financial currents: in the late 1970s, just as Morby was getting her start, changes in federal investment policy and tax law produced a markedly more hospitable environment for venture capital investment. First, the tax rate on the profit made from selling assets such as stocks or real estate, better known as the capital gains tax, was lowered from 35 to 28 percent in 1978 and then to 20 percent by 1980—part of the "trickle-down" theory of taxation intended to spur investment (or what Jimmy Carter's treasury secretary referred to as "the millionaire's relief act of 1978").[36] The overall supply of investment capital was also dramatically increased by a 1979 amendment to the Employee Retirement Income Security Act that made private industry pension funds available as a source of venture capital.[37]

In tandem, these transformations resulted in a surge of venture capital investment in new businesses; the change in the capital gains tax alone led to a fifteenfold increase in venture capital investment between 1977 and 1978.[38] And it was no coincidence that these changes

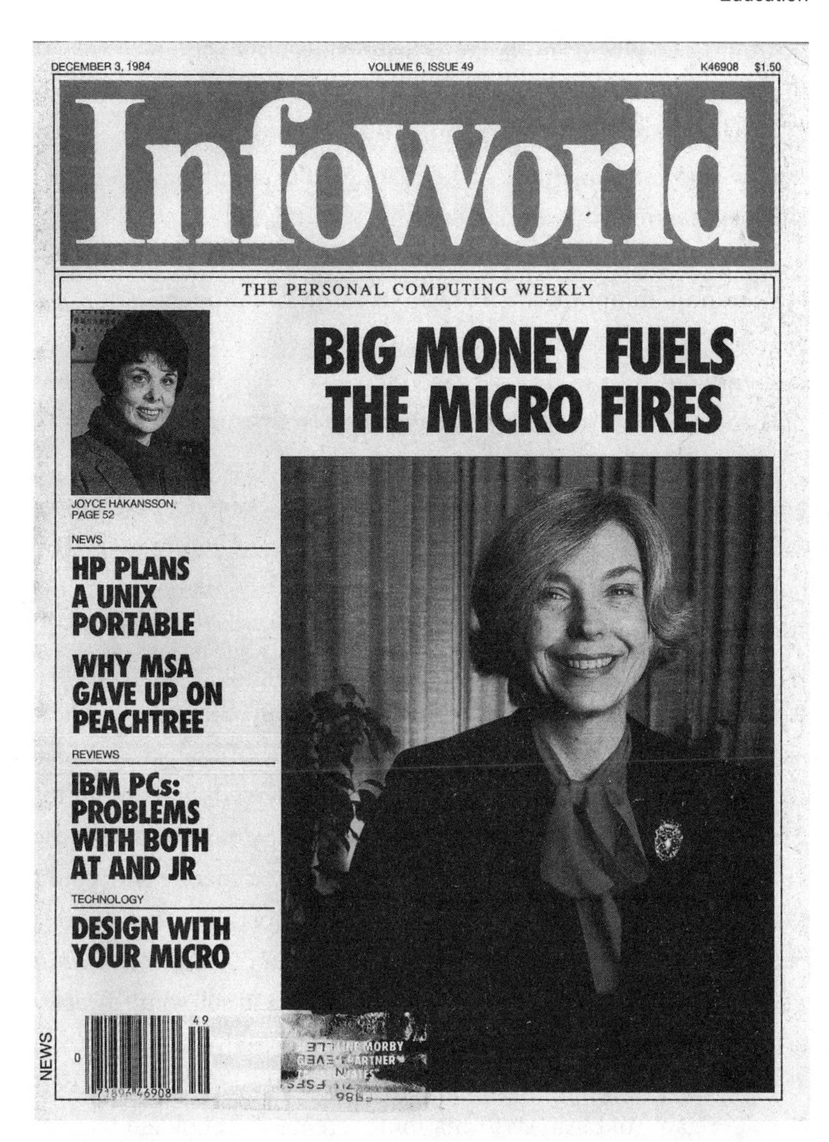

Venture capitalist Jacqueline (Jacqui) Morby, featured on the cover of *InfoWorld* for a story on venture capital in the microcomputing industry. December 3, 1984. Image from the collection of the author.

coincided with the rise of computing innovations and potential new markets: one of the chief lobbyists for changes to the capital gains tax was the American Electronics Association (originally founded by David Packard, of Hewlett-Packard), and even Robert Noyce, founder of Intel, testified to Congress on behalf of a capital gains reduction.[39]

The interests of venture capitalist firms converged with the interests of the high-risk computing industry to create a financial environment optimized for entrepreneurial activity. By the time Morby was sitting down with Bowman, then, she had a long list of computing prospects she was scouting as potential investments.

Morby was curious about Bowman's background in both marketing and minicomputing, but at her table, on her time, micros ruled the day. "While talking about microcomputers," Bowman recalls, "we both agreed that as prices went down the channels of distribution would change radically. They would go from specialized stores to mass retailers."[40] In other words, they both believed that microcomputing was at a turning point between its niche hobbyist past and its future as a mass technology. From their perspective, success would be held by whatever companies most effectively dominated these new supply chains, not necessarily by whoever had the best product to sell. But Morby also knew that few if any microcomputer software companies were prepared to handle such a transition at the level of their executive structure, as the upper management of many consumer software companies was still largely staffed by friends and family, regardless of training or experience.[41] When Morby narrowed her eyes on Bowman, then, some part of her surely saw someone who could skip the fuss, someone who would have no nostalgia for the cozy olden days of 1979 or 1980, when a programmer could just walk into a computer shop with a box full of floppy disks and sell some directly to the owner, or hire the kid down the block to sketch ad art. Bowman and Morby were birds of a feather insofar as they both understood the meta-game of business formation: to create shareholder revenue.

So Morby pitched Bowman on the spot, suggesting that he and Seuss develop a plan to more closely examine the microcomputer software industry. Three weeks later, just a few days before Christmas, TA gave Bowman and Seuss funding to conduct a three-week industry study to explore their most advantageous avenue for market entry.[42] Their initial assumption was that they would go into business software, but when they pursued retailers about their needs, a different answer arose: "We went to retailers and asked them 'What is the software that people ask you for and you don't have?' And the answer was always 'Educational software.' So we didn't arrive at our strategy through any brilliant analytical technique. We just asked retailers what the public wanted."[43]

But just what did this mean, back in the early weeks of 1982? What kind of "educating" did people imagine computers could do? The fascination with deploying computer technology for educational purposes had stimulated a wide variety of research initiatives throughout the 1960s and 1970s. This includes not just the previously discussed educational time-sharing systems that arose in computing hotspots nationwide but also creations like the computer scientist Seymour Papert's development of the child-friendly programming language LOGO, the computer-assisted instruction system PLATO, and a variety of experimental and behavioral psychology research initiatives that used computers.[44] Such ivory tower explorations did not necessarily inform entrepreneurial software development on the ground, however, nor was this research particularly well positioned to capitalize on the abrupt consumer interest in microcomputing. Founders of early educational software companies were typically, like Tom Snyder, teachers or instructional technology experts who had been bitten by the microcomputer bug; they paid some attention to pedagogic technique, but they were also limited in what they could accomplish given microcomputing's technical constraints.[45] Much of the software on offer simply added layers of graphics, feedback, or other simple interaction to standard educational activities based on rote memorization, techniques that played best to subjects in math, science, geography, spelling, and reading comprehension.[46] Designing software that could address more complex educational philosophies was difficult when consumer microcomputers still used slow-loading cassette software, and thoughtful approaches to design were not conducive to swift software release. Educational software developers and publishers thus had tremendous leeway in defining what qualified as "educational," and in many cases the sheer presence of the computer, applied to standard learning exercises, was enough to excite children and teachers alike.

Yet despite the fact that an educational software industry had begun to blossom by the early 1980s, it had not done so in a way that was particularly accessible to nonhobbyists, technical novices, or anyone who didn't want to spend hours poring over computer enthusiast magazines—in other words, the very potential consumers that Morby and Bowman and many other investors and entrepreneurs and MBAs saw as essential to the industry's profitability. On one end, massive publishing incumbents like McGraw-Hill were investing in courseware exclusively for sale to schools and pricing it accordingly: Snyder's *Search Series* went for $180 a set, a price on par with the wider instructional materials market but well outside consumer reach. On the consumer side, however, it was much less clear who products were for and how consumers were supposed to assess and obtain them. Even among the smaller educational software startups, like Edu-Ware, Steketee Educational Software, The Learning Company, and Basics & Beyond, products were primarily intended for schools; consumer marketing was largely an afterthought to these early companies' focus on formal learning environments.[47] Certainly the average Apple II owner could have called up Edu-Ware to order a program for practicing algebra, but even knowing that Edu-Ware existed, around early 1982, would have required a significant time investment reading scant advertisements and following industry news.

As businessmen whose jobs had been to refine the growth and operations of traditional consumer goods, Bowman and Seuss saw this as a solvable problem. They brought a way of thinking about business that focused on how to build channels for distribution by transforming computer software into the kind of mass merchandise they had handled in their previous work—a mode of economic evaluation that set them far afield from most other software startup entrepreneurs, where the tendency was to develop product first and figure out the market later.[48] The opening paragraph of their business plan, presented to TA Associates on February 15, 1982, tightly laid out their intentions:

> The company will publish software in the home educational and sophisticated game segment of the microcomputer industry. . . .

> Software will initially be purchased from outside authors, and the company will add value in documentation, packaging, distribution, and retailer support. The company expects to invest heavily in advertising and a direct sales force, in order to establish a brand franchise, build strong retailer support, and create a competitive barrier in marketing.[49]

This wasn't the microcomputer software industry of just a few years earlier, when publishers like Fylstra were running a business distributing software out of their living room or the Williamses advertised *Mystery House* under a made-up, unincorporated name. The majority of US consumers still may have been uncertain about owning a computer, but investors like Morby were confident that business chops, and enough startup capital to choke out small competitors through massive marketing buys, could set them up for success—even when they had no idea what they were going to publish beyond the genre. In Morby's estimation, "[Bowman and Seuss] didn't especially know the software market as well as we did, but they had an expertise that nobody in the software business had and we felt we could introduce them to the right people."[50] From an investment perspective, then, Morby found competency with code secondary to competency in business. It didn't matter that Bowman and Seuss had no product to sell, no experience in the market, no firsthand background programming on microcomputers. What they had, instead, was a strategy, wrapped up in a business plan, and the not unreasonable assumption that traditional economic laws would hold, no matter how cutting-edge, high-tech, or innovative the technology in question became. At the end of the day, you were still selling a box on a shelf in a store.

By the middle of March 1982, TA Associates approved $800,000 of funding for Bowman and Seuss—half of a $1.6 million valuation for a company that didn't even have a name, let alone software to sell, and was only the second venture-backed microcomputer software firm in the area (the other being Mitch Kapor's behemoth Lotus).[51] They eventually dubbed the operation "Spinnaker," an homage to the sailing culture that suffused the Charles River, which split Boston from Cambridge and played host to some of the oldest sailing clubs in the country. A spinnaker is a large, three-pointed sail at the front of a racing yacht or large sailboat. When full of wind, it produces a notable visual effect, ballooning smoothly over the bow like the puffed-out chest of a bird. But spinnakers are useful only under certain wind conditions; a boat must be sailing *downwind*, or in the direction of the wind, for a spinnaker to be of use. The company's stylized, arched-triangle logo and strong-leaning type treatment reflected these visual cues, but also spoke to other themes: Bowman, Seuss, and their investors believed they were heading straight in the right direction, riding an advantageous, and inevitable, tailwind. To launch the company, then, all Bowman and Seuss needed was product—and Jacqui Morby, whose son was enrolled at Shady Hill School, knew just who to introduce them to.[52]

* * *

In the quick rush of years since Tom Snyder had sold *Search Series* to McGraw-Hill, he had succeeded in shaping his life around his passions: writing rock songs, drumming out code, teaching the occasional science and music class at Shady Hill. He even found someone to be the president of his company, Computer Learning Connection, so he could spend his time ignoring the money and developing his craft—"just being me, no limits."[53] Doing ports and bug fixes and upgrades on *Search Series* kept the royalties flowing, but by the beginning of 1982, he was scoping his next educational computing creation: a deductive reasoning mystery game that would become known as *Snooper Troops*.

It would be a computerized take on a classic concept, in which children would form a team of detectives tasked with solving a small-town mystery. Imagine Nancy Drew or the Hardy Boys, tech-ified with a computer database of suspects and a radio wristwatch. Like *Search Series*, he envisioned a game that pushed students to play together, sharing the labor of managing different information streams and working collaboratively to make their deductions. The premise extended the pedagogic tendencies of Snyder's previous output, wherein children moved from confusion to clarity by gathering and then analyzing together a variety of information. But *Snooper Troops* also represented important shifts of interest in the microcomputer as an information management tool: he intended the game to be played not only by a group but also across three separate microcomputers, possibly networked together, enabling each child to engage in independent clue gathering built around the same story. His ultimate intention was to sell the program to McGraw-Hill for use in the classroom, maintaining the advantageous relationship he had already built with his prized publisher.

And all this might have happened had Bill Bowman and David Seuss not come knocking, directed to Tom Snyder by Jacqui Morby. Snyder had written a custom piece of software for Morby's son to help him with writing and composition—demonstrating the kind of dedicated, tech-savvy creative pedagogy a parent was not likely to forget.[54] Snyder had no memory of Morby contacting him before or discussing opportunities with him, but she had clearly filed Snyder away in her own mental Rolodex, in accordance with his talents, to be plucked out when needed.

So Bowman and Seuss arrived sometime in April or May, in hot pursuit of a deal, as they needed a master disk of whatever software they were going to manufacture by mid-July (as Snyder recalled, "The desperation level was high").[55] This timeline was fueled by the founders' ignorance. They had initially planned to release the product in spring the following year but later discovered that the majority of software sales were done during the winter holiday. It would be a surprise to them, again, that software had to be finished by midsummer in order to complete all the physical manufacturing and distribution in time to have products on the shelves for the Christmas season. It was, without question, a ridiculously short and untenable timeline for software development.[56] So while it was important that Snyder's idea for the game seemed appealing, it was more important that the idea was already well fleshed out and that he had the experience to get the software completed without much oversight.

The terms of this deal, however, required one crucial alteration to the software: the game had to be playable by a single child. As Bowman and Seuss had outlined in their business plan, the company was focused on publishing software in the "home educational" segment. While that same document acknowledges an ambition to sell to both consumers and schools, their priority was products that could flex across both contexts. What this meant, in practice, was that Snyder could no longer assume a group of children at play in front of a computer; the dialogic properties of the game had to be rewired to allow for a single child's experience. As Snyder recalls:

> As soon as we sold *Snooper Troops* to Bill and Dave, I threw the whole thing out and started over again, because it was now going to be for a consumer market. None of what you'd call the social interface that I had always designed for was relevant anymore. This was the first time I was writing something for an individual kid. . . . So I had to get rid of all of the fun part for me, which was the consensus and shared knowledge aspect of this thing. I could build into it a recommendation that two kids do it together, [but] Bill and David said, "Yeah you can stress that if you want, but it's got to work with one person."[57]

The idea that software was an individual product, made lucrative through its sale to many, many individual households, was obviously not unique to Spinnaker. Yet the forced transition of Snyder's game design nonetheless demonstrates how pedagogic strategies were co-opted by an economic logic that required software itself to express the notion that computers were experienced individually, or at least by a unit no larger than an individual family. Whereas Snyder's pedagogic techniques focused on the computer as a tool for enabling rich discursive engagement *between* children, Spinnaker's economic model required simplifying those dynamics to a circuit between the child and the computer, a set of one-on-one interactions that replicated the socioeconomic logic of the appliance-driven nuclear family. This, of course, was a logic Bowman and Seuss considered completely natural, given their background as marketing consultants for consumer goods; to move units of anything, you had to imagine people as individual consumers, not as learning communities or networks of practice. While the computer itself may have been cast as a revolutionary educational tool, its economics were purely classical.

And despite Snyder's pedagogic preference for collective play and consensus modeling, he too embedded the logics of consumer economics into *Snooper Troops* at its core: he insisted on building *Snooper Troops* around a generic data model, which would allow him to make multiple games using the same core logic (what we might identify today as a proto–game engine).[58] Thus Snyder committed to making not one but two *Snooper Troops* games for Spinnaker: *Snooper Troops Case #1: The Granite Point Ghost* and *Snooper Troops Case #2: The Disappearing Dolphin*.[59] The two games are functionally identical in game-play mechanics and graphics. The exploratory space is a gridded network of roads, dotted at regular intervals with numbered houses and traversed by a vehicle controlled via keyboard (the aptly named "SnoopMobile"). The primary pedagogic and ludic activity is the retrieval, documentation, and assessment of information, obtained by various exploration mechanics and written down by the player by hand. Players literally drive door-to-door, interviewing suspects one question at a time (and sometimes, inexplicably, breaking into their homes and photographing clues). The goal of all this information gathering is to discern which characters' alibis are legitimate by cross-referencing clues.[60] The game was, in essence, a database with pictures on top. By creating a programmatic separation between the code that made the game run and the lines of textual game data that players used to solve the mystery, Snyder was able to exploit a production model in which the game code was stable and the data files were interchangeable. As Snyder recalled, "I think I didn't change a line of code."[61]

With the benefits of a data model (and the contract work of a couple of professional freelance writers), Snyder would finish his games without much fuss, delivering the master disks to Spinnaker by midsummer. But from there, the process was more involved than the typical publisher-developer relationship. The *Snooper Troops* educational products were going to be shunted through a marketing, branding,

Game screens from *Snooper Troops Case #1: The Granite Point Ghost* (1982), running in MAME emulation on the Internet Archive. Top image illustrates the fictional town players driving around to visit suspects and gather clues. Bottom image illustrates the process of interrogating a suspect at the suspect's home. Screen captures by the author.

and distribution machine buoyed by the deepest pockets ever to sell a floppy disk for $44.95—a machine focused not on Snyder's software individually but on framing Spinnaker as a soon-to-be titan of educational publishing.

* * *

Having a product, in and of itself, was not the goal. What Bowman and Seuss wanted was a product *line* on which to build their notoriety as a brand. Whereas most early microcomputer publishers released one or two products first and only developed more once they had received some initial success, Spinnaker's founders took success as a foregone conclusion—something they could generate *through* the kind of consumer confidence that could be instilled by systematized management of branding and distribution channels.

To achieve this, the company cofounders hustled up two more pieces of software for their launch lineup while simultaneously prepping for packaging and advertising. There would be a Mr. Potato Head–style program called *Facemaker*, where children could swap out features on a digital head, and a children's animation program called *Story Machine*, both developed by a West Coast firm called DesignWare.[62] To bind the four launch titles into a coherent product line, Bowman and Seuss invested heavily in branding, marketing, and design. They understood packaging as a vital component in this process, the actual physical object consumers would hold in a store. While most consumer-grade software during this time was sold in either lidded or tabbed cardboard boxes, Spinnaker's products came in distinctive plastic clamshell cases.[63] Opening like a book, these cases offered a thinner profile and crush-resistant exterior, contributing to a sturdier and more professional appearance than typical packaging. All the product booklets were designed to the same dimensions (5.75 × 7 inches), and, along with the 5.25-inch disk, they fit so snugly inside the molded plastic interior that the box contents didn't rattle around when picked up, moved, or shaken (unlike other consumer software at the time).[64] These kinds of design decisions required a degree of upfront planning most consumer-grade software companies had never practiced or had the startup capital to support. Spinnaker's design team also recognized that branding could exceed the simplicity of a logo; the product packaging, advertisements, and manuals deployed an earth-toned gradient border to create design consistency across all products and support materials.

Altogether, these qualities distinguished Spinnaker's products from their competitors. There was no gawky adolescent Spinnaker, no charmingly awkward stage where the principals were copying disks in their kitchen or typesetting their own ads. Even if Bowman and Seuss were entirely unfamiliar with the stresses of software development and manufacturing, their consumer-facing strategy never let it show.[65] They bought prominent ads in *Creative Computing*, *COMPUTE!*, *PC Magazine*, and *Softalk*, handsome color spreads foregrounding a product line with a commitment to learning through play and promising a computer experience "so much fun your kids will probably forget they're learning" (see pl. 17).[66] Meanwhile, Jacqui Morby seeded Bowman as an interviewee for numerous newspaper articles that summer and fall, letting journalists position him as an authority on consumer educational software rather than a guy with no pedagogic or consumer software background running a company that had yet to release a single product.

Spinnaker's products would land in September, optimally timed with both the start of a new school year and the seasonal runway into Christmas. *Snooper Troops*, for its part, was well received by reviewers, especially for its versatility in appealing to adults and children alike. "Parents who buy *Snooper Troops* for their kids will burn the midnight oil solving it after the kids are bedded down," applauded Margot Comstock Tommervik in her September 1982 *Softalk* review of the new program.[67] *Creative Computing* was similarly impressed with the game's depth, reporting that "although this is billed as an educational adventure for children, we found it was quite challenging and not something that could be easily solved in 10 or 15 minutes."[68] The sheer scale of information players were required to assess made *Snooper Troops* a game players had to truly work at, above and beyond the artifice of its educational mechanics. Billed as appropriate for ages "10–Adult," the game proved to have a breadth of playability across age groups that many other educational (and even straight-up adventure) games lacked.

While *Snooper Troops*' depth and versatility primed it for positive reception, Spinnaker's entire lineup benefited from entering the market at a time when industry efforts to *create* educational software were colliding with rising consumer interest *in* educational software. It was no small thing when the *New York Times* dedicated the spring 1982 edition of its triannual "Survey of Education" to the topic of computers in schools, with *Times* education editor, Edward B. Fiske, writing a substantial analysis of the computer's capacity to alter intellectual life.[69] Surveying education scholars and computing researchers ranging from the Nobel Prize–winner Herbert Simon to the computer scientist Alan Kay to the artificial intelligence researcher Allen Newell, Fiske's article leaves readers with the impression that the computer would not simply expand our reach on information, but "open up the possibility of entirely new forms of learning, teaching and thinking itself."[70] While the presence of computers at all levels of schooling had been on the rise in the previous few years, articles like Fiske's ensured that these trends would only amplify within an American public bracing itself for an alteration to its very way of being. This faith in the revolutionary capacities of computing was further disseminated by computing and education researchers such as Arthur Luehrmann and Seymour Papert, as well as the notable futurist Alvin Toffler, all of whom circulated their ideas across industry, academic, and mainstream publications.[71] As microcomputers increasingly became figured as technologies of a future everyday life, even the most watered-down version of these theories—like believing a child will gain distinct advantages just by being exposed to a computer—became a ground truth stimulating consumer assumptions about computing's educational power. This was precisely the zeitgeist Bowman and Seuss had stumbled upon in their retailer survey, when they assessed that software storefronts were receiving requests for "educational software" that barely yet existed in the early 1980s.

All this assertive emphasis on the impact of computing in education essentially became a feedback loop, spurring the growth of a consumer software sector that had ostensibly not previously existed. Much like the home software category, the rise of consumer-facing educational software as a distinct category was several years behind more prominent arenas like business, games, and utilities, and for similar reasons: early microcomputers like the Altair, and even the earliest second-generation systems of 1977, were not especially approachable to nonhobbyists, children especially. Such machines were expensive, complicated, and, in many ways, delicate, which did not lend them to being characterized as children's toys in the way equally complex but internally inaccessible video game consoles were. At this early stage, assumptions about the computer's ability to educate children were typically based on the presumed benefits of exposure to a computer rather than anything explicitly pedagogic about the computer or its software.[72]

Thus, while a smattering of educational software can be located in hobbyist magazine advertisements immediately following the release of the 1977 Trinity, the offerings were extremely thin and routinely categorized as "home" software.[73] This market grew erratically over the first years of the 1980s—especially because the term *educational* was doing a great deal of unacknowledged work. Magazine editors and reviewers routinely conflated any program with instructional properties as educational, regardless of the age range addressed, and expensive courseware intended for schools was often featured or reviewed alongside products priced for general consumers.[74] For example, a two-part educational software roundup published across the September and October 1980 issues of *Creative Computing* not only included software *for* teachers (in other words, industry-specific workplace programs) but also evinced the kind of jumbling common to this software category, in which products like Personal Software's educational cassette program about birth control were listed alongside RadioShack's grammar- and middle-school math drills. The premise for what was considered educational was flexible if not self-referential. To select products for review, the article's author, David

Lubar, defined "educational" software using two criteria: "those labeled 'Educational' by their manufacturers" and "those which, while not labeled 'Educational,' do provide the user with new concepts, new information, or new approaches to problem-solving."[75] Categorization, then, contains its own industry logics: the first key to producing educational software was to simply declare it "educational."

In an explicitly Apple context, the question of an educational software category first received significant attention in *Softalk*'s December 1982 issue, coinciding with the addition of two Spinnaker products to the Apple user magazine's specialized best sellers lists: *Snooper Troops Case #1* and *Facemaker* (reflecting sales for October that year).[76] Neither carved its way into the magazine's Top Thirty, but each made a strong enough sales showing that *Softalk* was obligated to list both programs somewhere: *Snooper Troops Case #1* placed third in the "Fantasy 5," tying for bronze with the popular role-playing game *Ultima*. *Facemaker* ranked eighth in *Softalk*'s "Home 10," reflecting earlier tendencies in software ads to position educational software as an extension of domestic activity. Classifying *Snooper Troops* as a fantasy game appeared to be a contrivance for the sake of expediency; while acknowledging that "*Snooper Troops I* will probably get relegated to a new educational list soon," the editors felt the game operated within the framework of a "children's fantasy role-playing game."[77] In a note that underscored the novelty of educational software, *Softalk* acknowledged that *Facemaker* was one of only two pieces of "educationally oriented software ever to score on the Home 10" (the other was the child-friendly programming language *Apple LOGO*).

Thus the Christmas season of 1982 proved to be the tipping point for educational software's consumer arrival—a set of circumstances Spinnaker's very existence no doubt contributed to. *Softalk* would launch a "Home Education 10" the very next month, in January 1983, as a way of capturing the full breadth of November 1982's holiday shopping. As the editors wrote:

> Nearly twice as much software was sold in November as in October, and not all of it was entertainment. What might best be called soft

> education packages are making a big showing. Soft education packages are those that sugar-coat the learning process, generally are not curriculum-based, and are dependent more on home buyers than on school adoptions for sales. . . . In recognition of the changing circumstances in educational software, *Softalk* has implemented a Home Education 10 category.[78]

The "Home" in "Home Education" was an important signifier in the context of *Softalk*'s best-sellers listings, simultaneously pointing to both a place of use *and* a type of consumer. *Softalk* was not in a position to track curriculum-based sales that went through specialized educational distributors, a fact they openly admitted in this issue. Here a market gets defined, honed, sharpened against itself: while it was certainly the case that home education products could make their way to the classroom, or vice versa, the individual consumer, rather than the student user, is the only trackable entity.

By parceling out home education as its own category and asking for reports from retailers on what they were selling in that space, new products came into view and others were recategorized, offering an abrupt index on this slice of the software market. The inaugural edition of Home Education 10 was led by Lightning Software's *MasterType*, a typing instructional game (a similar though less gamified product, Microsoft's *Typing Tutor*, also made the list). Both programs were categorized as educational by virtue of their instructional orientation, both largely targeted adults, and both had been extracted from the "Home" category they had previously dominated.[79] While having remarkable market longevity (*MasterType*, in particular, would remain an education best seller for years), these programs were anomalies in the larger lineup of "sugar-coated" software, 70 percent of which was devoted to children and most of which had never been on any *Softalk* best-seller lists.[80] This included two Apple-published educational games developed by the Children's Television Workshop, the producers of *Sesame Street*, as well as *Step By Step* and *Early Games for Young Children*, both one-off programs from small developers.

Over the course of 1983, educational products increasingly cemented themselves as top thirty best sellers among the overall Apple software industry. By November 1983, *Softalk* declared that "the educational arena has been the area of fastest sales growth and has become the most competitive of all. Where you once couldn't get a computer retailer to carry anything educational, the product list of educational products actually selling in any significant volume is now longer than any other category."[81] In less than a decade, the microcomputer's educational potential, once merely an article of faith, had become a winning economic proposition—even if its actual utility at home or in the classroom had yet to be proven.

* * *

Spinnaker's inclusion among the *Softalk* best sellers was fairly erratic. *Story Machine* was the first program to post on the Top Thirty, in March 1983, but only for that month (likely reflecting January 1983 purchases from families that bought a computer for Christmas); *Facemaker* appeared in May 1983 and again in July. And *Snooper Troops*, for all its praise, did not hit the Apple II best-seller list until June and July 1983, ranking twenty-seventh and seventeenth, respectively. However, this level of fluctuation is not surprising given the density of the Apple II market overall, the steady performance of Spinnaker's preexisting competitors, and a number of new arrivals to the home education scene during this time, including *Rocky's Boots* from The Learning Company and Xerox's *Stickybear* line of games for young children.

Yet while Spinnaker may have rarely topped the charts with individual products, its strength was in its *spread*, in how many products consistently did well enough. For example, while five different companies commanded the top five rankings of the Home Education 10 in September 1983, Spinnaker monopolized the entire bottom half. This generally reflects a strong position within retailers: Bowman and Seuss understood that the retail game was all about how many inches of shelf space a company could command. Because they were offering a product line, the performance of any one product was less important than the performance of the line overall. Thus even if the company's individual products sold less than tentpoles like *MasterType* or *Early Games*, the company was more broadly diversified by virtue of its product mix, especially as it continued with new releases such as the LOGO-like *Delta Drawing*, the preschooler-friendly program *Kindercomp*, and a new Tom Snyder adventure game, *In Search of the Most Amazing Thing*.

Yet Spinnaker's true strength lay beyond the Apple market. By 1982, there were too many platforms on the market for software developers to be wedded to a single hardware platform. Whereas software was once the tail wagging the hardware dog, these dynamics had shifted rapidly, as developers and publishers had to increasingly make strategic decisions about which platforms to serve. Spinnaker's initial 1982 product launch was compatible with the Apple II, the IBM PC, and the Atari 400/800 (disk version only), indicating Spinnaker's assumption that educational software was chiefly a high-end users market. By 1983, however, Spinnaker expanded to manufacturing cartridge versions of its products, for both the Atari 400/800 and, more important, the Commodore 64. Released in 1982, the Commodore 64 razed the low-end computing market across 1983 through aggressive price cutting, rising from 23 to 50 percent of the home computer market by the middle of that year.[82] Commodore's price competitiveness accelerated industry assumptions about the inevitability of computer adoption. Meanwhile, countless companies, Spinnaker included, followed Commodore's lead by releasing cartridges for a platform many thought would make the microcomputer a mainstay family appliance.

Moreover, aware that Commodore had a mass-marketing angle that would put it in stores like Sears and Kmart, Spinnaker activated a strategy that played to its cofounders' strengths. Bowman and Seuss knew how to aggressively approach mainstream retailers, and they had a product line that looked ready for mass marketing. As Seuss recalled:

> A vice president at the Boston Consulting Group used to say, "big boys play with big boys and little boys play with little boys." And we

> acted like and dealt with the big boys. We immediately went after the largest retailers, looking for the best market share. Our background really helped in this thinking big.[83]

In order to drive consumer awareness of Spinnaker's product in stores, the company campaigned aggressively during the Christmas season of 1983, becoming the first educational software company to buy ads in mainstream consumer magazines like *Better Homes & Gardens*, *Good Housekeeping*, and *Newsweek*.[84]

These fall 1983 advertisements are an object lesson in Spinnaker's approach to an expanded market (see pl. 18). Whereas Spinnaker's original ad framed the company's products as learning games children *wanted* to play, as opposed to more tedious, rote, or poorly developed educational software products, the 1983 ad campaign configures the software as "computer games you want your kids to play," directly addressing parental suspicion about whether something was educational or just a game. These advertisements assume an intended consumer that lacks computer savvy, mostly associating any kind of interactive, colorful computer activity with the pointlessness of video games. To counteract those assumptions (or, as Spinnaker puts it in the final sentence, to "giv[e] computer games a good name"), Spinnaker's ad campaign places its software packages within the grasp of the child, who is in turn under the explicit guidance of two parents.[85] Visually, both advertisements summon a set of class and race anxieties almost identical to those circulating in the home software market, like those seen on the cover of the *Family Computing* launch issue that same fall (as discussed in chapter 6). However, rather than configure the computer as part of the family as a whole, these ads center the child. The parents are cut off; only the whiteness of their hands appears, awkwardly resting on their child's shoulders. The mother wears a wedding ring, while what we can make out of the clothing—the gold-buttoned dark blazer, the knit sweater, the lace-edged sleeve—tells us everything we need to know about the class position of these parents. Here the computer becomes an extension of parental guidance and authority, something to be exerted as gently but consistently as a guiding hand on the shoulder.

Between its expanded marketing and the quartet of consumer computing it served, Spinnaker thrived. While it closed its first fiscal year in January 1983 with sales of around $750,000 (an impressive number given that Spinnaker had only been selling product for five months), it closed its 1983 fiscal year with sales of an expectation-shattering $11.3 million—a 1,400 percent increase in revenue growth and roughly three times what the business plan anticipated.[86] This astounding growth is best reflected in Spinnaker's performance on *Billboard*'s Education Top 10 software charts, which tracked software sales across multiple microcomputer systems.[87] From when *Billboard* first started tracking software releases in October 1983 through spring 1984, Spinnaker never had fewer than four products on the best-seller list; its dominance peaked the week of February 11, 1984, when seven of the titles on *Billboard*'s Education Top 10 were Spinnaker software.[88]

Spinnaker's economic vitality would prove the rule of venture capital: Jacqui Morby and TA Associates didn't want to own half of a $1.6 million company; they wanted to own a smaller percentage of a much larger company. Spinnaker was revalued at $9.5 million in December 1982, then ratcheted its valuation to $50 million less than a year later, in October 1983.[89] That round was a global investment effort, including General Electric's pension fund, private investors from Europe, and continued investment from current stakeholders.[90] As one industry analysis put it, "This was probably the highest valuation ever given to a company of Spinnaker's characteristics."[91]

Spinnaker was supposed to go public in 1984—and why not? Business was booming, and Spinnaker had strategies. There were plans to grow into adventure games, arcade games, and home productivity software. To stave off the threat of incursion from traditional toy manufacturers like Fisher-Price, Spinnaker pursued partnerships instead.[92] To push its brand recognition further, the company made deals to sell products in Walden Books stores. As individual educational products aged out, Spinnaker came up with a better idea than throwing inventory in a dumping ground: it invented a bargain bin brand, renamed its products, and sold them as $10 games.

And as for the country, educational anxiety had never been higher. In addition to the general drumbeat of educational computing boosterism, a 1983 federal report titled *A Nation at Risk: The Imperative for Educational Reform* gave renewed substance to the argument that the United States was in dire educational straits, unprepared for the "knowledge, learning, information, and skilled intelligence" that were the "new raw materials of international commerce."[93] If the prospect for US schools seemed bleak, home education software would be a stopgap for parents who could afford it—all of which was good for the industry and for Spinnaker, until it wasn't.

* * *

Spinnaker learned the same lesson as every other microcomputing company that year: there is no such thing as endless growth. The downturn was due in part to a new phase of competition within the education market, as what *Softalk* described as "old-line companies" in both computing and publishing made incursions into the education market, driving a software bubble that popped about as badly as the games market.[94] The larger piece was, of course, the industry at large—which had confounded consumers with too many hardware options and continued to offer poor rationales for consumer spending. But it wasn't as if Spinnaker had stopped making money. Rather, it was that they didn't make as much money as they thought they would, relative to how fast they were spending it. But by the time they realized growth was slowing, in mid-1984, it was too late to throw the brakes entirely. Spinnaker's sales grew to $15 million, but its market share stayed the same. By some estimates, Spinnaker was the leading independent software publisher in the country, and on par with Brøderbund in 1984. But the market was not growing at a rate that affirmed the company's exuberant valuation. It turns out that so-called expert sales forecasts largely served only to *inspire* investment rather than actually *predict* it—a sleight of hand aiming to create growth as an aftereffect of claiming growth would happen.

For a company of Spinnaker's size and strategic investment, however, the industry downturn was something of an advantage as long as they were able to weather the storm. The drying up of ready venture capital meant few new competitors to contend with once the company emerged into 1985, with a readjusted expectation of $19 million in sales. Bowman sloughed off and out of company leadership in 1986, as part of a readjustment to the reality, rather than simply the potential, of personal computing. As Seuss reported to the *Boston Globe*, "Both Bill and I started Spinnaker, we expected 40 to 50 percent penetration of computers in the home. . . . We expected to build a big company and that hasn't happened because computers in the home are in only four percent of American homes."[95] That reality meant that both the founders and Spinnaker's primary investor, TA Associates, had "yet to hit it big by selling shares to the public."[96] Soon thereafter the company would pivot out of education to focus on small business and personal productivity software. For Bowman, Seuss, and Morby, this had never been about an essential truth concerning the value of computing in education. Everyone here was an opportunist of their moment. Spinnaker made educational software because retailers said they wanted educational software. When that market fell short of its promise, Spinnaker's founders and private shareholders pressed to move on.

Tom Snyder, for his part, reverted to making educational software for school-based learning, as well as a broader range of creative business pursuits that sheltered under Tom Snyder Productions. Snyder continued this quiet, fairly anonymous work for a decade, until his company was picked up by the Canadian mass media corporation Torstar in 1996 (a company notable, according to the *Boston Globe* article announcing the merger, as the "owners of Harlequin romance novels").[97] While nearly 80 percent of Tom Snyder Productions' revenue came from educational software and the school market, the remaining 20 percent came from a recently launched enterprise that remains, even today, probably what Snyder is best known for: the creation of the "Squigglevision" animation technique, most prominently deployed in the animated series Snyder co-created, *Dr. Katz, Professional Therapist*, which debuted on Comedy Central in 1995.[98]

And in the long, strange arc of this world, Tom Snyder Productions would come all the way back around. The company was bought by the educational publisher Scholastic in 2001, somewhere to the tune of $9 million, where the acquired company's strength in interactive educational software was expected to help buoy Scholastic through the latest wave of predictions about the necessity of educational computing in the nascent Internet Age.[99]

That same year, in 2001, a Stanford education professor, Larry Cuban, published *Oversold and Underused: Computers in the Classroom*. The book was a scathing evaluation of two decades of computer use in schools. In Cuban's estimation, while a wide coalition of stakeholders—including "public officials, corporate executives, vendors, policymakers, and parents"—had proffered reform agendas for increasing accessibility to new technology in schools, the nearly religious faith in the computer's inherent capacity to drive economic productivity had made computing in the classroom little more than "an expensive, narrowly conceived innovation."[100]

What Cuban's work dramatically underscores is how correct companies like Apple had been to seek their payday on the back of American educational aspirations. Apple would maintain the strong position in the education market it had claimed in the early 1980s, especially following the release of the Apple Macintosh. With its uncomplicated graphical user interface, novel mouse peripheral, and a chassis that literally couldn't be opened (rendering it antihobbyist but increasingly childproof), the Macintosh was extraordinarily successful at feeling *like* the future, even if there was no real apparatus in place within the educational system to make it so. The struggles of the US education system were, and remain, intractable, deeply baked problems around which there is no clear consensus. In the absence of cooperative political will, glowing screens and plastic keyboards are the lowest possible hanging fruit—and one that, about every decade or so, we seem to forget we've already eaten.

Inconclusions

To end, let's go back to the beginning. Back to Steve Jobs's 2007 iPhone keynote.

The iPhone keynote was a one hour nineteen minute event, packed with turns of phrase that could seed a new generation of dissertations: "touch your music"; "visual voice mail"; "the killer app is making calls!" Some of the most important internet and telecommunication CEOs of the era show up as guests to walk across stage and pump Jobs's hand: Eric Schmidt (CEO of Google three CEOs ago), "Chief Yahoo!" Jerry Yang, and Stan Sigman, head of Cingular. There's something so easy but unflattering about these appearances, men confident yet graceless as they trot across stage, pace in their chinos, babble about the power of partnership. "We're going to bring some great stuff to market over the years together," chirps Jobs near the end. These are financial mating rituals.

In the keynote's closing minutes, Jobs—ever the showman, ever the storyteller of himself—returns to the historical set pieces that opened his talk. He must remind the audience of this technological inheritance that is theirs by right. The words *Mac* and *iPod* appear again on the monumental slideshow that forms Jobs's only backdrop, followed by the logo for Apple TV, then the word *iPhone*. Jobs narrates over the slide transitions:

> So, today, we've added to the Mac and the iPod. We've added Apple TV and now iPhone. And you know, the Mac is really the only one that you think of as a computer. Right? And so we've thought about this and we thought, you know, maybe our name should reflect this a little bit more than it does. So we're announcing today we're dropping the "computer" from our name, and from this day forward, we're going to be known as Apple Inc., to reflect the product mix that we have today.

The screen behind Jobs illustrates the disappearance. One moment, the slide reads, "Apple Computer, Inc." And with the flick of an animation command, the word *Computer* is blown away into particles, like a spritz of condensed air casting dust from a keyboard. Apple might make computerized devices, but Jobs saw no reason to tie those devices to a particular *form* of computing. What Jobs intuitively recognized was that to keep the company growing, Apple needed a grander story from which to rationalize new technological investment—because, really, where else did it have to go? In this way, Jobs recasts the story of Apple as if the computer itself was only ever a step on the way to something else, as if that had been the intention all along. In Jobs's new mythology, the computer is everywhere and nowhere. It's in your pocket, it's on the wall; it's a box on a desk, it's a foregone conclusion. For Steve Jobs, the turn from personal computing to ubiquitous computing was to be manufactured as self-evident and inevitable. But he can only get there by naturalizing personal computing to begin with, by not asking the question of how it happened, what forces allowed it to emerge.

This book, in contrast, has taken those questions as its starting point. How did these machines become *possible*, and why did they *matter* to people at *this* moment in time? What we have walked through together across these pages is a story operating at many scales. First, the microcomputer arrived as a convergence of preexisting technologies and tendencies within computing, and the communities formed around its use. You needed a mode of data processing based on the construction of a user, achievements in electronics miniaturization that were decades in the making, a robust network of early adopter hobbyist users fixated on the power that came with technological ownership, and immediate runways for profit-making in order to scale production. All these phenomena converged in the 1970s, resulting in the emergence of the first consumer microcomputers like the Altair 8800, and then by 1977, the oft-mentioned Trinity of the Apple II, the TRS-80, and the Commodore PET. Steve Wozniak and Steve Jobs's development of the Apple II was a thoughtful blend of open engineering and marketing savvy that won over investors, hobbyists, and tech novices alike. The software market that exploded around the Apple II thus left behind a wealth of documentation allowing for a reconstruction of the force of impact with almost geologic precision.

The software histories that unfolded from that point are unified here in their narrative progression. They reveal how microcomputing practices began as somewhat indeterminate and grew clearer as a variety of stakeholders jockeyed over computing's possibilities, potential, and future. Yet each specific piece of software covered—from famed programs like *VisiCalc* and *The Print Shop* to long-forgotten oddities such as *Locksmith* and *Snooper Troops*—has given us a window into the unique particulars of software development, publishing, marketing, and reception. By leveraging an individual piece of software to discuss the larger body of a software category, each chapter has offered a centralizing account while also documenting a more generalizable domain of use, telling us something of how the earliest modes of American consumer computer practices emerged.

But throughout it all, we can clearly see the intractable role financial speculation and the construction of markets played in people's desire to even imagine what shape innovation might take. The technological itch and the creative instinct do not preexist the economic conditions in which they arrive. In the hands of the early adopters profiled here, microcomputers were simultaneously a vector for their creative exploration and wealth accumulation. Whatever Dan Bricklin's fantasies of shredding numbers with lasers shot from an X-wing, he and Bob Frankston fundamentally understood themselves as men trying to get paid. In the many turns of *VisiCalc*'s development, they scoped and molded the software to fit the constraints of existing technical and economic conditions, while their publisher Dan Fylstra marketed the product by speaking directly to financial anxieties buffeting the nation. Roberta Williams may have designed her game at her kitchen table, but she did it in the larger context of her husband's insatiable late-night grind as a freelance programmer, a rapacious energy they shared and which drove them as a couple. *Locksmith* may have been designed as a programmer's experiment, but it took no effort for the author, whoever they were, to foresee its commercial opportunity—and set the entire Apple II world ablaze around the question of what the "right" to do business even means. David Balsam and Martin Kahn were perfectly positioned at a successful, well-resourced company like Brøderbund to put their creative talents together and figure out how to get a piece of the action pulsing through the Bay Area. Tom Snyder had always been trying to make a living doing the things he loved, from music to teaching to making toys and games for kids. Like the engine model he designed for *Snooper Troops*, the anticipation of selling was built into the making.

Yet these software acts were not, in and of themselves, enough to make personal computing personal. As this book has demonstrated, the microcomputer had to be sold—not just on shelves, but in the hearts and minds of potential users. A tremendous amount of social and financial effort went into naturalizing and normalizing the notion of a computer of one's own. It was not simply that users had to reimagine these technologies in order to make them practical or useful, though examples of such behavior abounds in this era. It was that personal computing emerged welded to an affective imagination that sutured individual creativity, discovery, and opportunism to a sensation of national rejuvenation via technological innovation. This is the source of the electric energy and earnest capital reinvestment that kept the consumer microcomputing industry afloat, no matter how many times it was proved that mass consumer interest was tenuous, uncertain, or sometimes not there at all. It is in the hope of the thing that might come that every time personal computing failed, it was forgiven and allowed to try again—until computing's embeddedness in our world *did* become a foregone conclusion.

So did everyone get what they wanted? Did you? We should be doubtful. As we know from how this history turned out, removing the computer as the center of Apple's mission paved the way for its comparatively recent reliance on digital services, rather than digital products, to foster increased growth and shareholder value. The present-day ubiquity of digital services, of computer devices we never turn off, of networks and signals ever humming, ensure that companies like Apple are ever ready to capture more "value" in the form of our time and attention, our hardware upgrades, our service contracts, our content subscriptions. The turn to a consumer computing market ruled by digital services is one in which the computer increasingly disappears from view, just like it did in Jobs's keynote, even as the operations it manages encroach ever further into our lives. If we want to claim, as so many do, that computing's past is an energetic and inspirational resource for future generations, then perhaps we should bother to *know* that history—the ambiguity of computing's actual origins, the rapacious indifference of the industry's rampant financial speculation, the struggle most people had to make anything of it—and begin to dismantle, for ourselves, the histories given to us by those most invested in our compliance with a notion of inevitably escalating technological progress.

EPILOGUE

On the Consignment Floor

In October 1912, an essay ran in the radio engineering trade journal *The Marconigraph*, telling of a "wireless girdle round the earth" currently under development by the Marconi Wireless Telegraph Company of America.[1] Written by the company's chief engineer, the essay outlined how a chain of high-powered transmitter and receiver stations would allow wireless signals to hop and skip from one fulcrum of empire to the next—from Panama to Hawaii to Manila, from Bangalore to Aden and onward to Europe, "cross[ing] the wide Mediterranean, ascend[ing] the boot of Italy, scal[ing] the ice-crowned Alps and drop[ping] quietly into London, all in less than one two-thousandths of a minute."[2] To complete the circuit, however, required a receiving point on the eastern coast of North America that could pass the signal across the great expanse of the continent and back out to the Pacific. And as it turns out, the Marconi Company had purchased just such a site that very year, ensuring the wireless girdle would cinch itself together "near New York City, at Belmar, N.J., where 500 acres of land have been acquired upon which the masts and plants will be erected."[3]

A century and nine years later to the month, in October 2021, I stood on a small patch of that long-ago purchased land, looking out at the two-and-a-half-story red brick building that had once operated as the Marconi Hotel (this was "hotel" in the archaic sense, as a lodging house for the radio operators who manned the receiving station twenty-four hours a day). In the intervening years, this building, and the ones surrounding it, had lived many lives: commandeered by the US Navy during World War I; serving as a summer resort for the Ku Klux Klan during the interwar period and, briefly, home to a nondenominational Christian college; transforming into Camp Evans during World War II, as the US Army cannibalized the preexisting telecommunications infrastructure and established the base as a national center for military radar and electronics research.

Where I imagine there had once been a series of military checkpoints, only chain-link fencing and overgrown foliage shrouded the entry. Protected as a National Historic Landmark, the property operates today largely as a memory palace of what it once was, playing home to the InfoAge Science and History Museums ("New Jersey's Mini-Smithsonian!!," as the online tour map declares), a collective of regional history initiatives operating as a STEM-focused educational nonprofit.[4] These include the New Jersey Shipwreck Museum; a World War II homefront re-created living room; collections of military communications technology, radar, and signal electronics; World War II battlefield dioramas; model train installations; an "African American History Room" focused on Black male engineers and scientists; a room somehow simultaneously dedicated to the "Lenni-Lenape American Indians," European colonization, *and* the Revolutionary War; the Military Technology Museum of New Jersey, which seemed to be, most visibly, a squad of tanks sitting in the sun; and something called, in the brochure at least, "Fallout Shelter Theater."[5] It was InfoAge that had brought me to this scratch of land along the marshy southern edge of New Jersey's Shark River tidal basin that Saturday afternoon. Or more specifically, what was happening *inside*

InfoAge that Saturday: it was the second day of the Vintage Computer Festival East, an annual regional retrocomputing convention—the long, long, long tail of the hobbyist collectives that once thrived around early microcomputers like the Apple II.[6]

Billed by the program as a "fantastic family-friendly adventure backward in time," the festival is a certain kind of window into a certain kind of imagination about a certain understanding of our technological past. Instructional sessions offer deep dives into topics that could have been pulled from the headlines of 1970s microcomputer magazines: CRT repair; Atari 2600 graphics programming; mastering the abacus and the slide rule. A slate of keynotes ran over the weekend, featuring names and companies you might now recognize from having almost finished this book: David Ahl and Ted Nelson; Michael Tomczyk of Commodore and Scott Adams of Adventure International. This is the history of computing largely reduced to its technical function, a world where most of what has changed between 1977 and today is the specifications of hardware: the amount of memory, the speed of processors, the channels available for audio, the number of sprites, the catalog of colors, the shift in relative cost. Materiality is also a key component of the appeal, going beyond simple notions of look and feel. We're talking about the springiness of a keyboard, the resistance of a joystick, the sound a disk drive makes when it spins, which can tell you if it's in good working condition or not. Honoring the material legacy of these machines, a makerspace hosted a variety of hardware kits for tinkering and exploration, as well as an introduction to soldering. Meanwhile, exhibits comprised the majority of what there was to interact with. These were the pride and joy of the festival, where local retrocomputing enthusiasts set up tables to show off their hacks, mods, and historic systems in a fashion reminiscent of the first microcomputer fairs of the 1970s and 1980s—or perhaps the way Steve Wozniak once stood by a card table in the hallway outside the Stanford auditorium where the Homebrew Computer Club met and dared passersby to take an interest in an obscure little 6502 computer board.

I was accompanied (and perhaps chaperoned) on this adventure by

Signage at the InfoAge Science and History Museum. Computer history is situated in the same context as US military history and hobbyist electronics exhibits. Photo taken by the author, October 9, 2021.

Jason Scott, a former UNIX systems administrator turned computer history documentarian turned internet celebrity cat owner turned resident mad hatter of the Internet Archive—the man largely, almost solely, responsible for coordinating the Internet Archive's extensive compilation of historical materials related to US computing history, as well as emulated software from the 1970s and 1980s (a collection of materials to which this book is indebted for its very existence).[7] Scott had driven me for over an hour out of Manhattan and down New Jersey's Garden State Parkway, spinning yarns most of the way about the endlessly convoluted tangle of personalities that constituted the formative history of the festival and its organizers. Together, we roamed the grounds more like gonzo ethnographers than actual participants. Scott, with his wiry beard and silhouette-defining top hat, stalked the event through the lens of his massive DSLR camera. I, clad in a neon green jumpsuit, one of the very few female-bodied people there, occupied myself by trying to soak up the grounds, running through and cutting across a spartan array of vaguely similar buildings composed of seemingly identical hallways, doorways, and meeting rooms occasionally converging at 90-degree turns, like some deeply uninspired *Doom* mod. We played an original *Pong* machine; lapped the elaborate holdings of the New Jersey Antique Radio Club's Radio Technology Museum; bought some actual 1970s computing convention T-shirts and vintage copies of *Creative Computing*; chatted with exhibitors; got bored, wandered off, climbed on tanks. Scott's phone blew up with texts the minute he came on campus. No one knew, or particularly cared, who I was.

One of the reasons I had gone to the festival on Saturday was to hear a talk by David Ahl, the *Creative Computing* founder whose observations about computers and public opinion opened chapter 1. But I could barely make it through a quarter of his more than hour-long lecture about the first fifty years of video game history. As Ahl tabbed through some PowerPoint slides on computer history's greatest hits—the codebreaking functions of Bletchley Park, Alan Turing's imitation game, Joseph Weizenbaum's 1950s chatbot ELIZA—I fussily texted Scott: "How does anyone in this room not already know these things?" His reply was a thesis nearing haiku:

Everyone knows these things
This is church, you heretic

I had made the mistake of coming to the altar of a god I don't worship. As a historian, I attend something like the Vintage Computer Festival not to celebrate this past, consecrate my devotion, or engage in the ceremony of remembrance but to understand the historical argument such communities make about what the past *is*—an endeavor that involves attention to the textures of both speech and silence. There is, as Scott suggested, a *faith* to everything that happens here, a belief that this past must be honored, acknowledged, remembered, saved, preserved—though to what exact ends are unclear. Much like InfoAge itself, the Festival takes the educational value of what it does as self-evident, not because any of the technologies on display are still in use today, but because they constitute (so the story goes) a step in a great chain of innovation that is believed to express humanity's function on this earth. This assumed faith in the operations of historical technological ancestry is made clear in InfoAge's mission statement:

"to preserve, educate, and honor scientific innovation to inspire new generations of thinkers, dreamers, and visionaries."

What all this amounts to is a certain kind of *orientation* to the past, to what makes history meaningful. There is something so patently ridiculous about imbuing tanks and radar guns, H-bombs and wireless stations, and, yes, even computers, no matter how personal, with a notion of innocence divorced from their conditions of production, their material application, and their economic circulation. And yet this is what so much computing history as written does, over and over again—demanding both to be celebrated as the latest entry in a tremendous line of technological extension, transfer, surveillance, and broadcast, and, simultaneously, to set aside the political, financial, and social violence of that history by forever resurrecting a case for its own virtue through its popularization as a mass consumer plaything.

A primary goal of this book has been to offer a history that exceeds the heroic identification with and technical mystery surrounding what were, in fact, *products*. This is not to diminish them—capitalism exerts its own affective powers—but to understand that the dominant role computing plays in our everyday life might have more to do with conquest, control, and, especially, capital than freedom, creativity, or anything resembling progress. American culture treats financial success as the reward for technological innovation, accepting that the tech barons of the world have earned their astronomical, mathematically incomprehensible paydays. But what if it were the other way around: what if technological "innovation" is just the other side of market anxiety, a stress response, and computer technology just a pliant medium for siphoning capital from one site to another? Every couple decades, for the past fifty years, consumer computing technology has rewired, somewhat, how the financial pipelines flow, and who they flow to. But it has never, ever, changed the fact that the flow moves from the many to the few.

This became most evident where our tour concluded: at the festival's dedicated consignment floor, a massive open space under gut renovation. Consignment floors are staples of retrocomputing events, a place where small-time dealers and overstocked collectors create a kind of communal flea market, a bargain bin of computational afterlives. I took several laps, trying to feel the room. It was the southeastern wing of one of Camp Evans's two "H-Complexes," built during World War II to house the Camp's radar production operations. The floor was smooth, unforgiving concrete; the walls and ceiling appeared to be unfinished drywall; intense fluorescents ignited the pure whiteness of the space. Insulation poked out from the exposed window frames, mingled with fuzzy spiderwebs that seemed much larger than they should have been allowed to grow if anyone was maintaining the space. If you looked carefully, you noticed moments of broken glass in the window panes.

At the farthest edge of the room was overflow seating for the lectures, where visitors could sit and watch the talks via projector; the unevenly toned audio bounced chaotically off the unfinished architecture. Beyond that, however, the space was dominated by an array of wooden tables with folding steel legs, the kind used at cheap, large-scale catering events. And I remember how *empty* it looked. So little was for sale in consignment that everything had been spread out across the entire room, yet this only served to heighten the sense of emptiness, making the space feel monotonous and abandoned—like they had set up tables for a wedding of five hundred, but only a few dozen of the party guests ever arrived. COVID was part of that, surely—though Scott insisted he had never seen the festival so packed—but it felt melancholic to me in a different way, like every object was on leave from the Island of Misfit Computer Equipment.

I wandered the tables while Scott bantered with the merchants in charge. There was little rhyme or reason to the layout. A box full of CueCats here. A tangled pile of ethernet cables over there. A bin of 64 KB core memory boards. Ancient, untouched rolls of pin-feed paper, $10. A boxed Tomy Tutor, LaserJet printer cartridges, a nest of tech conference swag, with a handwritten sign that simply read, "FREE LANYARDS." And everywhere I looked: Apple products. I was drawn to an artfully haphazard arrangement of four Apple II disk

The sparsely occupied consignment floor at the 2021 Vintage Computing Festival East, held at InfoAge. Photos taken by the author, October 9, 2021.

drives atop a beaten rectangular table, cables unbound, prices marked in black Sharpie on blue painter's tape: "VCF $30." Not even eight bucks in 1979. (See pl. 19.)

Apple's detritus piles up. Several tables down, a squad of mid-aught MacBooks and PowerBooks, 20-some-inch external monitors, and an iMac G5 sat undisturbed in the angular shadows of some cool, mid-Atlantic afternoon light. Behind me were iMac G4s, their earnest screens craned slightly upward on sleek mechanical necks, blocky old ImageWriter IIs and a Personal LaserWriter, "Graphite" PowerMac G4s in their distinctive plastic chassis, slot-loading iMac G3s (color:

Blueberry). All around me was a history of consumer computer design that was alleged to have changed the world—*Insanely Great*, as the tech journalist Steven Levy cast it in his book about how the Macintosh "changed everything"—much of which could now be yours for under $100.

Here history dies, even as it allegedly lives on. Everything this book has traced—the longer arc of innovations and cultural practice that made microcomputing possible, the technical travails of Wozniak, the erratic showmanship of Jobs, the curious spark of hobbyists, the untethered risk-taking of the market's earliest entrepreneurs, the uncompromising self-interest of that same market's earliest investors, the unquantifiable *work* an entire subculture put into making these allegedly magical machines legible and useful to a population of consumers more often feeling anxiety than inspiration, the way economic crashes and geopolitical competition conspired to underscore the necessity of a computer-literate nation—all of it somehow got left behind here. I've said it best elsewhere: "History is not in what we talk about, but in *how we organize its meaning*."[8] The Vintage Computer Festival, InfoAge, entire arms of pop journalism, YouTube videos, and the publishing industry are all occupied with telling us the same

Outdated, unwanted Apple products scattered throughout the consignment floor of the 2021 Vintage Computing Festival East. Photo by the author, October 9, 2021.

nostalgic tale about computing: how it inspires, the futures it will make. Our willingness to consume these narratives is directly proportional to our own desire to see ourselves in history—what I sometimes imagine as a secular version of ancestor worship, or one of the few culturally acceptable spaces for white people to fixate on legacy, now that appeals to the legacy of whiteness itself are increasingly socially inappropriate.

Throughout the process of writing and editing this book, my developmental editor, Tim Laurio, asked me, with every chapter, to assert what I meant about what made the computer "personal." He nudged this out of me with careful comments, massaged sentences, restructured conclusions. And I yielded to this line of inquiry, because, after all, it's the title of the book. But in the end, there is no grand synthesis that tells us what, through some range of theoretical and academically acceptable justifications, I think the "personal" is. Honestly, my aims have been more feral than that, and simpler: this has been a heist, tailored to rob as many people as possible of their much-cherished faith in computing's primordial innocence by showing how compromised, fraught, and indifferent, to all of us, this history actually is. Disrupting nostalgia is a dangerous game but worthwhile. It forces a reckoning with why our fantasies of history take the shape they do, a shape always more informed by the present than whatever is lost to the past.

This is what I like about the consignment floor: here everything seems to become more honest. Visitors fawn over these same technologies in the exhibit halls, where they are carefully assembled, copiously documented, arrayed to put on a show. Here they're just something to trade for fiat currency. Stripped of their purpose, scattered about in defiance of any natural chronology or implied trajectory of influence, everything is entirely forgettable, just waiting to be packed back up in a plastic box if they can't be sold. Everything is waiting to live again, even though most of it can't, won't. Here we tell the truth: there sits a floppy disk drive, once a marvel of modern technology, created by what remains one of the wealthiest companies in the world. And you can barely give it away.

Acknowledgments

I documented the progression of this book in a single continuous Twitter thread that began on January 7, 2020. For the subsequent twenty-eight months, I posted to that thread after nearly every day I worked on the book, transcribing the amount of time I had written, my chapter word count, and a small comment on my progression, struggles, and general emotional state. I chipped away at this book in pockets of time that could measure thirty-five minutes or more than three hours, but that thread documents much more than pure chronology. It carries the imprint of a world that shattered beneath all of us. Scrolling through the hundreds of tweets that comprise the thread, I am transported across a mosaic of experiences I never intended it to capture: the unvarnished panic of early COVID lockdown; the unanticipated exhaustion of all work becoming from-home work; the deprivation from human contact; the protests, the marches, the choppers beating overhead, the sirens that wailed against silence. Tucked between the folds of some tweets are dramas personal to myself: falling in love, breaking up, chronic pain, getting older. And the slow emergence of alternative patterns of life: the outdoor hangouts and editing sessions, the new attention to weather, "returning to the office," fantastical regimes of sanitation theater—and then, somehow, the way it all just stopped.

But what that thread affirms to me, more than anything, is the extent to which this book was written *in community*—regardless of whether the people in my life ever liked a tweet or knew the thread existed. After all, if it wasn't Twitter it was something else: Zoom, WhatsApp and Signal groups, daily Slack posts, Instagram DMs, SMS, good old-fashioned phone calls, and gobs and gobs of email. To every person mentioned here, and many more I'm sure I've forgotten, you have played a role, made a contribution, been aid to me along the way. This book is indebted to all of you.

There are three people, specifically, without whom this book would have remained a shadow of its initial potential. First, to my editor at the University of Chicago Press, Joseph Calamia. Your warm patience and thoughtful feedback pushed this book to live up to its own intentions. Then, to my agent, Sarah Levitt. Years ago, you saw something in me that I had not yet seen within myself, and that has made all the difference. Last, the secret hero of this book is undoubtedly my developmental editor, Tim Laurio. If any reader finds this book well structured and competently written, that praise belongs to him. Tim, you brought the keenest eye and the deepest respect to this work; so often your small, tender margin comments made me smile and kept the fire burning on this work more times than you realize.

Beyond those who pushed this book forward are those responsible for its genesis. This book would not exist without strategic mentorship from my senior colleagues in the Department of Media, Culture, and Communication at NYU, namely, Finn Brunton, Sue Murray, Natasha Schüll, and Marita Sturken. This was not the book I had set out to write when I arrived at MCC, but I am so indelibly grateful that it's the book I wound up writing. Thanks, to each of you (though perhaps Marita especially), for being as stubborn as I am.

To my colleagues at MCC, and the NYU/NYC academic scene,

more broadly: individual contributions are hard to enumerate, but a few special mentions are warranted. My appreciation, again, to Sue Murray, who aside from being my department chair throughout most of the writing process, has been a reliable, considerate mentor and probably my biggest cheerleader. Mara Mills, also, has been an incredible source of mentorship, collegiality, and leftover CSA food. To my fellow junior colleagues, Whitney (Whit) Pow, james Siguru Wahutu, and Angela Xiao Wu, thank you for the encouragement, trust, and respect. Likewise, this work would not be possible without the help of MCC's talented staff and administrators, who aided me in countless matters related to work and teaching; thank you to Rebecca Blough, Darrell Carter, Tracy Figueroa, Annette Morales, Dove Pedlosky, Dani Resto, Carlisa Robinson, and Winnie Wu. Intellectual and personal support also often exceeded the bounds of the departmental. My thanks to the industrious band of scholars, thinkers, and makers of games and art across NYC, especially Joost van Dreunen, R. Luke DuBois, Colleen Macklin, Nancy Nowack, and John Sharp.

I finished this book in the context of teaching my first doctoral class at MCC, Computing History. Class conversations were top of mind as I pulled together the final edits on this manuscript. I would like to thank that class for their dedication, thoughtfulness, patience, and curiosity: Tony Brave, Ria Citrin, Chelsea Dan, Shikhar Goel, Nabil Hassein, Grace Lim, Charli Muller, Jinhee Park, Lane Yates, and Nansong Zhou.

In terms of tracking down stray materials and historical leads, I was assisted greatly by Henry Lowood at Stanford University Libraries and David Brock at the Computer History Museum. Much of the historical basis for this work comes from the computer enthusiast magazine *Softalk*, which I was fortunate to have explored early in my career thanks to help from folks at the Softalk Apple Project. Furthermore, I was able to explore the *Softalk* archive in great depth during my time teaching graduate classes at Georgia Tech, especially while working with Kera Allen. Our early conversations about the history of both *Softalk* and *VisiCalc* informed my thinking on those topics; thank you to Kera for being such a thoughtful interlocutor. I would also like to thank the anonymous scholars who peer reviewed the manuscript, whose thoughtful feedback improved its final form.

For assistance with images, I'm indebted to Massimo Petrozzi at the Computer History Museum; Alex Cranz for corporate leads; Kevin Driscoll for fair use help; Jenni Balthrop at Hewlett-Packard; Kay Peterson at the National Museum of American History Archives Center; Beth Merkle at the Strong Museum of Play; Brad Herbert at the Sierra Museum; Andrew Borman and Tega Brain for last-minute photoshoots; Andrew Berger for finding the exact photo at the exact moment I needed it; Steve Wozniak and Dan Sokol for proving the sharing spirit of homebrew still lives on; and Mary Eisenhart, Pat Johnson, Jason Scott, and Efrem Sigel for help with last-minute image permissions.

I am phenomenally appreciative to the individuals who gave their time to be interviewed for this book: David Balsam, Doug Carlston, Martin Kahn, Corey Kosak, Tom Snyder, and Ken and Roberta Williams (who had previously been interviewed for other projects). I think that telling one's own story is easy; being entrusted to tell someone else's is a far harder task. For those of you who figure especially prominently in this book—namely, David, Marty, and Tom—I hope what I have written has captured, to some small extent, something of the spirit of the time and your experiences.

Jason Scott (@textfiles), "resident mad hatter of the Internet Archive," as I refer to him in this book, deserves his own paragraph, so by god I'll give him one. Jason, you went above and beyond in your dedication to helping me with this book—serving, at various turns, as an unofficial research assistant, an intrepid social media amplifier, and an always affirming internet dad. I hope it makes you proud.

As the opening to these acknowledgments suggests, my relationships on Twitter, many of which are only on Twitter, have composed a kind of ambient background static against which this book has been written. The sense that there were people "out there" who cared about its content meant I never felt alone in writing. So my thanks go out to every reply guy, every "like" and retweet, every person who stopped to humor my technical questions or chattered about archive images. Some even gave me leads or helped me understand the tough technical problems early microcomputing is full of. I kept a list of helpers as I went (with apologies to anyone I overlooked): @apple2europlus, @brouhaha, @bzotto, @cjmenning, @dosnostalgic, @enf, @foone, @JeffAStephenson, @kaysavetz, @n1ckfg, @sehugg, and @ultramagnus_tcv. In addition, @a2_4am generously provided extensive insight and crucial fact-checking on my discussion of copy protection schemes.

I have received support, encouragement, and occasional "you got this!" memes from several colleagues beyond my home institution, including Lori Emerson, Matthew Kirschenbaum, Soraya Murray, Lisa Nakamura, and especially Patrick McCray, who read early versions of the earliest chapters. I also owe a special thanks to Ian Bogost, for his early support and feedback on this project. In addition, I'm grateful to the communities I have worked with and learned from at both the Special Interest Group for Computing, Information, and Society (SIGCIS) and *ROMchip: A Journal of Game Histories*, especially Morgan Ames, Peter Sachs Collopy, Gerardo Con Díaz, Stephanie Dick, Kevin Driscoll, Jason Gallo, Raiford Guins, Xiaochang Li, Henry Lowood, Andrew Meade McGee, Soroya Murray, David Parisi, Elizabeth Petrick, Joy Lisi Rankin, Andy Russell, and Melanie Swalwell.

Now we get to the hard part: tracing the networks of friendship, the lay lines of care that make work possible even if the contribution is not quantifiable. First, to the Media Slackers: Jacob Gaboury, my work husband in all things; Carlin Wing, for gripping authenticity; Dave Parisi, the indefatigable; and Stephanie Boluk and Patrick LeMieux, for believing in—and modeling—the life well played. For boats, bilges, and bourbon, friendship forged by moonlit rafts and reading 'round the fire: N. D. Austin, Josh Bisker, Danielle Butler, and Celeste LeCompte. For antes, all-ins, and letting me take your money, more often than not: William A. Anderson, Alex Cranz, Tim Hwang, Rachel Lovinger, and Rachel Mercer. To Tega Brain, Sam Lavigne, and Chris Xu—y'all are a silver lining on the darkest timeline. Ramsey Nasser, for rides, and Stephanie Dick, for rituals. Kevin Cancienne and Margaret Robertson: you are family, full stop. For much-needed respites of reading outside, noodle dinners, and Xena binges: Whitney (Whit) Pow and Jessie Roy. Charles Eppley: I miss you. To Celeste (again) and Erin L. Thompson, for gifting me queer community. Liz Barry, you slid into my life just as this book's final edits closed—a surprise and a delight of the highest order. Five-star reviews and endless thanks to my physical therapist, Lizz Lemontagne, without whom I quite literally could not have written this book. To Brent Strang and AC Deger: I should have called more.

And then, there are two people in the world who carry legal responsibility if I ever go on life support: Finn Brunton and Kette Thomas. Finn, you are a friend out of legend, a fantasy book hero, with a beard destined for Valhalla. You are as bright as the sun, and you obliterate my shadows. And to Kette: your friendship defines the mortifying ordeal of being known. I owe you everything.

A Note on Archives and Sources

A word on sources and archives is pertinent here, as well as timing (even historians can't escape it). This book was conceived in fall 2019; four months later, in spring 2020, the COVID-19 pandemic dramatically altered, and in many cases halted, the movement of bodies across states, nations, and oceans. The initial months of the pandemic also brought with it new apparitions of economic crisis, causing strident lockdowns on university research accounts and travel funds. I was fortunate, in conceiving this book, to have organized it around a corpus of material I was already fairly familiar with and had readily at hand: the Computer Magazines Archives housed on the Internet Archive, a nonprofit digital library that holds literally billions of items, including over 28 million books and texts and nearly 600,000 software programs (as of January 2022). The Computer Magazines Archives is itself a collection of smaller collections totaling close to 30,000 periodicals (though some are duplicate scans). While not encompassing every computing magazine ever published, these collections capture a great deal of the computing enthusiast and industry magazines that formed the early basis of news and knowledge for microcomputer owners, including *BYTE*, *Creative Computing*, *COMPUTE!*, *Family Computing*, and especially *Softalk*, which was the only magazine to run rigorous, monthly best-seller charts within the software industry, making it a critical archive for this work.

Beyond the magazine collections, the Internet Archive also hosts scans of many other valuable materials, such as hobbyist newsletters, marketing materials, operating manuals, and the like. The existence of these works on the Internet Archive is a testament to the devotion and goodwill of a small, furiously dedicated community of retrocomputing enthusiasts to individually scan, or coordinate the bulk scanning of, thousands of pieces of ephemera. The OCRing of these texts as part of their existence on the Internet Archive, allowing for quick keyword searching of tens of thousands of pages of content, also permitted this project to be undertaken at a speed and scale that would've been impossible just a decade or two before.

In addition to the materials housed on the Internet Archive, this work made use of my own eclectic collection of early personal computing guide books, manuals, buyer's guides, encyclopedias, and dictionaries (once I could get back into my office to access them), as well as more traditional sources like major newspapers and national US periodicals, which I was able to access online. When necessary, I supplemented primary documentation with personal interviews over phone or Zoom.

For much of the writing of this book, it was not possible to travel to archives. Had the option been available to me, I likely would've reviewed the Apple-related holdings at the Stanford University Archives and the Computer History Museum. However, only chapter 2 is explicitly about the history of Apple Computer (rather than the broader computer software industry). Given that this was the chapter I was writing exactly at the moment COVID hit New York City, I designed the chapter to fall within the scope of digitally available primary and secondary documents. The book overall does not attempt to make original claims about Apple's specific company history, nor does it need to for its larger themes to bear out. Even as travel became more plausible during various phases in the (still ongoing) pandemic, I ultimately determined that the Internet Archive's breadth had allowed me a unique, and perhaps unprecedented, form of due diligence to the subject matter herein.

Notes

Introduction

1. For articles that reported directly from Jobs's iPhone launch, see Block, "Live from Macworld 2007"; Honan, "Apple Unveils iPhone."
2. Clips of Jobs's presentation are widely available on YouTube, though the keynote is worth viewing in its entirety; see Jobs, "iPhone 1."
3. "The Shakeout in Software."
4. For discussion of the claim that computing ownership was only in the double digits by the mid-1980s, see chap. 6, n. 3. For information on the installed base of personal computing in the United States in the 1990s, see Eric C. Newberger, "Computer Use in the United States," *Current Population Reports*, prepared by the US Department of Commerce and the US Census Bureau, October 1997, issued September 1999; Eric C. Newberger, "Home Computers and Internet Use in the United States: August 2000," *Current Population Reports*, prepared by the US Department of Commerce and the US Census Bureau, issued September 2001.
5. This lineage is traced in Bardini, *Bootstrapping*. The media theorist Anne Friedberg presents largely the same history in the final chapter of her monograph, *The Virtual Window*, framing personal computing as a primarily visual medium.
6. For primary document references on this subject, see Coburn, *Learning about Microcomputers*, 1–42; Willis and Miller, *Computers for Everybody*.
7. In this book, I have made the decision to italicize the titles of all commercially released software products, which is a departure from most bibliographic standards, including Chicago style. Chicago style mandates that software and applications

should not be italicized. Games are the sole exception to this rule; according to the seventeenth edition of the *Chicago Manual of Style*, games should be italicized "like movies, a usage that recognizes the narrative and audiovisual similarities between the two art forms" (8.190). Yet in a book about the dawn of consumer software as both a cultural and an economic form, insisting on a bibliographic distinction between games and other software products is, in this historian's opinion, untenable. Chicago style's rules rely on an outdated analogy between games and other types of moving-image media, and, furthermore, determining what even constitutes a computer game is not self-evident. Microcomputer game software was manufactured, published, marketed, and purchased through the same channels of production and distribution as all other microcomputer software at this time. In keeping with the larger political economy approach of this book, I elect to treat all packaged, individually sold software products as published works. However, this categorization does not apply to software released for free or circulated by homebrew collectives (such as TINY BASIC), operating systems (like Apple's floppy disk drive DOS 3.3), or programming languages and interpreters distributed with original hardware (the Apple II's Integer BASIC and Applesoft BASIC, for example, both stored on ROM). When quoting from any primary documents that do not italicize the title, I have left the quotations in their original typographic formatting. My appreciation to the media historian and archivist Peter Sachs Collopy for helping me think through these particulars.

8. I set aside two other significant components of Apple II software history as outside the scope of this book: word processors and the operating system CP/M. Word processing has a complex history in the larger scope of computing development and office automation. As Matthew Kirschenbaum documents in his monograph, *Track Changes*, word processing changed the labor of composition itself in both material and conceptual ways. While word processors are extremely relevant to the emergence of personal computing, I elected (largely due to constraints of space) to focus on software categories that had not yet received monograph-level attention. As for the disk operating system CP/M (an abbreviation of Control Program/Microprocessor), it materialized as a hardware peripheral, Microsoft's Softcard, which enhanced the software library of the Apple II. Softcard and CP/M, however, did not constitute a category of software in the ways conceptualized in this book. For more on word processing, see Haigh and Ceruzzi, *A New History*, 208–13; Kirschenbaum, *Track Changes*. For an explanation of CP/M and Softcard, see Weyhrich, *Sophistication*, 133–36.
9. For a very general history of the Macintosh, see Levy, *Insanely Great*.

Chapter 1

1. For comprehensive histories of the development of computing in the United States, see Campbell-Kelly et al., *Computer*; Haigh and Ceruzzi, *A New History*.
2. Ahl, "Computer Power," 46–47.
3. Ahl, 42.
4. Campbell-Kelly et al., *Computer*, 241. For a critique of the 1977 mythology as it pertains to the public awareness of personal computing, see Nooney, Driscoll, and Allen, "From Programming to Products." If there is a moment that more accurately reflects when computing reached the American public, it is likely found somewhere

between *Time* magazine's January 3, 1983, issue, which declared the personal computer "Machine of the Year," and the now-legendary 1984 Macintosh Super Bowl ad.

5. The definition of a computer is not stable, and varied historical interpretations exist. However, such questions are beyond the scope of this book. While many mechanical and electromechanical "computing" devices existed prior to the 1940s, I borrow from Paul E. Ceruzzi's designation that "modern computing" in the United States "developed after 1945 in a climate of prosperity and a strong consumer market. It was also during the Cold War with the Soviet Union." *Modern Computing*, 7.
6. Footage from this night, including several moments of interaction between the commentator Charles Collingwood and the UNIVAC, can be found on YouTube. See CBS News, "Election Coverage."
7. Such representations and concerns were coherent with general domestic anxieties of the late 1950s and early 1960s around automation and labor—John F. Kennedy's establishment of the Office of Automation and Manpower in 1961 being just one example.
8. For more on the history of mainframe computing in the postwar United States, see Campbell-Kelly et al., *Computer*, chaps. 4–7; Ceruzzi, *Modern Computing*, chaps. 1–5.
9. Estimate drawn from Rankin, *People's History*, 13.
10. Adjusted for inflation, these values translate to roughly $22,220 to $1,022,300 in 2020 dollars. Numbers are based on the monthly leasing cost of the IBM 360/20 and the IBM 360/67 as of summer 1968. See "Monthly Computer Census."
11. Rankin provides an effective narrative close-up of the steps involved in accessing a mainframe and the various constraints of these types of systems. Rankin, *People's History*, 12–15. For an account of gendered labor in the history of American computing, see Abbate, *Recoding Gender*.
12. This explanation of time-sharing comes from Ceruzzi, *Modern Computing*, 154–55. Also see Campbell-Kelly et al., *Computer*, 203–5; Wilkes, *Time-Sharing*. While time-sharing becomes commercially and institutionally popular in the late 1960s, its precedents can be found in the real-time interactivity first piloted in large-scale mainframe computing installations like SABRE and SAGE (used for airline reservations and military airspace monitoring, respectively).
13. Driscoll, "Professional Work," 259.
14. Terminals, including teletypes, were usually not computers in their own right; they lacked memory and independent processing power. Rather, they were designed with the telecommunications capacities to serve as remote input/output devices (thus they were sometimes called "dumb terminals," on the premise that they had no computer "brains" of their own). For a more detailed discussion of terminals and teletypes, see Kirschenbaum, *Track Changes*, 124.
15. The term *minicomputer*, while generally applicable, was not always used. IBM tended to refer to its minicomputers as "midrange" systems. Minicomputer setups could also be referred to as "small business computers" or just "small systems," depending on their marketing and implementation. Distinctions between mainframes and minicomputers could be somewhat arbitrary, especially once all systems began using microprocessors. For examinations of what makes a minicomputer a minicomputer, see Campbell-Kelly et al., *Computer*, 216–18; Ceruzzi, *Modern Computing*, 124–39. For a case study of minicomputing in educational contexts and its relationship to the broader history of personal computing, see Rankin, *People's History*, chap. 3.

16. Campbell-Kelly et al., *Computer*, 216.
17. Ceruzzi, *Modern Computing*, 133.
18. Ceruzzi, 124.
19. Rankin, *People's History*, 4–5.
20. The early to mid-1980s saw a flourishing of instructional and educational materials intended to stimulate the interest of children and adults alike in computing. See Litterick, *How Computers Work*, 10; Conniffe, *Computer Dictionary*, 64; Coburn, *Learning about Microcomputers*, 14.
21. Statistic drawn from Campbell-Kelly et al., *Computer*, 223. The role of US military interests in fueling miniaturization research is well documented. See Ceruzzi, *Modern Computing*, chap. 6; Lojek, *Semiconductor Engineering*, esp. chap. 7; Misa, "Military Needs."
22. None of these technologies represents a clean break with the technology before it. The transitions between standardized electrical components were diffuse, and there was typically a decade or more of overlap between them. For the history of the integrated circuit, see Campbell-Kelly et al., *Computer*, 215–25; Ceruzzi, *Modern Computing*, chap. 6; Kilby, "Integrated Circuit"; Lojek, *Semiconductor Engineering*.
23. For a contemporary account of the rise of semiconductor manufacturing in the United States, see Vacroux, "Microcomputers."
24. For a brief summary of Moore's law, see Ceruzzi, *Modern Computing*, 217. Footnote 43 from that section provides a useful account of common misinterpretations of Moore's law, as well as direct references to Gordon Moore's 1975 paper on the subject. Moore's earliest examination of the concept, however, was in 1965; see Moore, "Cramming."
25. The microprocessor is a good example of simultaneous invention. Several companies were working on such a technology in the early 1970s, although Intel is widely regarded as the most important. The story of Intel's development of the 4004 illustrates many of the technology's core economic and technological properties. See Campbell-Kelly et al., *Computer*, 231–32; Ceruzzi, *Modern Computing*, 217–21; Faggin et al., "History of the 4004"; Noyce and Hoff, "Microprocessor Development."
26. Quote from Vacroux, "Microcomputers," 32. Michael Orme refers to this scaling as the "learning curve" of the semiconductor industry. See Orme, *Micros*, 127–28.
27. Orme, *Micros*, 128.
28. How did a general-purpose microprocessor like the 4004 "know" that it was supposed to perform the functions of a calculator? The 4004 microprocessor was just one part of a set of four chips Intel released that allowed it to operate like a computer. Instructions for any given implementation were stored on a ROM program memory chip (the 4001), which could be more easily customized and cheaply produced. See Faggin et al., "History of the 4004," 16. Together, these four integrated circuits (which included RAM and input/output registers in addition to the ROM and the CPU) were, in Intel's language, a "micro computer system." This terminology comes from the original ad that Intel ran in the November 15, 1971, issue of *Electronic News*.
29. Olsen's alleged quote has been widely circulated, though its provenance is unclear. Olsen was cited making two similar statements: "I can't see any reason that anyone would want a computer of his own" and "There is no reason for any individual to have a computer in his home." The first quote is credited to a May 1974 meeting at DEC, reported by Ahl, "Dave Tells Ahl," 72. The second is cited to an address Olsen gave to the World Future Society in 1977, which is reported to have been republished

in *Time*. Olsen has modified and qualified this statement on numerous subsequent occasions. See Mikkelson, "Ken Olsen."

30. These limitations made Intel reluctant to announce the 4004 for commercial sale because they believed that "customers accustomed to the power of minicomputers would be unable to adapt to the microprocessor's poor performance." Faggin et al., "History of the 4004," 18.

31. Ceruzzi, *Modern Computing*, 244–55.

32. Ceruzzi, 222.

33. The wide-ranging character of hobbies has resulted in a comparatively small literature on the subject. Kristen Haring's *Ham Radio* is definitive among those works on technical hobbies. I specify "electronics hobbyists" as a subset of what Haring terms "technical hobbyists," which includes activities like motorcycle modding or model airplane construction. Haring defines a technical hobby as requiring "some technical understanding or skill beyond simply how to operate a technology." See Haring, *Ham Radio*, 2. While sheer numbers are impossible to know given the nature of hobbies, *Popular Electronics*, one of the most popular hobbyist magazines, had a distribution of more than 433,000 in 1974, equal to about 1.6 percent of US households. Annual circulation and distribution numbers are listed in the back of each January issue from 1962 on. *Popular Electronics*, January 1975, 110.

34. Haring, *Ham Radio*, 76.

35. Examples drawn from *Popular Electronics* cover features in January 1971, July 1971, October 1971, and November 1971. The November 1971 article on a build-it-yourself electronic calculator was written by Ed Roberts, cofounder of the electronics hobbyist manufacturer Micro Instrumentation Telemetry Systems (MITS)—the company that would release the Altair 8800 microcomputer in 1975.

36. Exceptions prove the rule. The earliest known computing hobbyist group dates to 1966, when an engineer, Steven Gray, founded the Amateur Computer Society (ACS). The group's first newsletter had 160 subscribers and encouraged the circulation of information on building computers. Nearly all subscribers would have worked in computing in universities, defense firms, or the computing industry itself. Campbell-Kelly et al., *Computer*, 218–19. For analysis of the ACS newsletters, see Gotkin, "When Computers Were Amateur."

37. Haring, *Ham Radio*, xii.

38. Haring, 75. It is estimated that anywhere from 40 to 73 percent of technical hobbyists held jobs in engineering and science, compared to only 2 percent of all Americans of working age. Haring, 81.

39. Haring, 157.

40. Haring, 78–88.

41. For an examination of the broader masculinization of the computer programming industry beginning in the 1960s, see Ensmenger, *Computer Boys*, 77–79.

42. Haring, *Ham Radio*, 44–48. For information on African Americans working in the electronics field, see Ford, *Think Black*; McIlwain, *Black Software*; Shetterly, *Hidden Figures*.

43. Both the Scelbi-8H and the Mark-8 used the Intel 8008, the company's second commercial microprocessor. Ceruzzi, *Modern Computing*, 225. For the original article on the Mark-8, see Titus, "Build the Mark 8."

44. Helmers, "What Is BYTE," 6. Influence could also go the other way: in that same issue, one subscriber and ham radio operator wrote, "Amateur radio is a natural for the computer hobbyist." Campbell, letter to the editor, 87.

45. For an account of the founding of MITS, Ed Roberts's role, and how the Altair wound up on the cover of *Popular Electronics*, see Freiberger and Swaine, *Fire in the Valley*, 27–36.

46. The $397 price would equal about $1,900 adjusted for 2020 inflation. There was also a stripped-down version for $298 and an assembled version for $498. Roberts and Yates, "Altair 8800," 34.
47. See Ceruzzi, *Modern History*, 227–28, for an explanation of the Altair's technical advantages over predecessors such as the Mark-8 and Scelbi-H8. One of its major advantages was MITS's decision to implement the 8-bit Intel 8080 microprocessor, which had just been introduced in 1974. While the cost of using the latest microprocessor should have priced the Altair well beyond hobbyist reach, the Altair's designer, Ed Roberts, leveraged Intel's inexperience with the hobbyist market to get a significantly reduced price—allowing the Altair to be first to market as technically superior to earlier kit computers *and* comparatively inexpensive.
48. The politicization of technical exploration was core to the Hacker Ethic in the US, which espoused, "access to computers . . . should be unlimited and total," "all information should be free," and "mistrust authority—promote decentralization." See Levy, *Hackers*, 39–51.
49. For more on the legacy of radio kits and the tension between building and buying, see Haring, *Ham Radio*, 49–73.
50. Salsberg, editorial, 4; my emphasis. While the Mark-8 and the Scelbi-8H had technically accomplished this in 1974, the Altair's use of the more advanced Intel 8080 microprocessor made it substantially more minicomputer-like than those prior systems.
51. For further explanation of this historical interpretation, see the epilogue of Joy Lisi Rankin's *A People's History*, titled "From Personal Computing to Personal Computers."
52. Numerous books document the creative and technical communities that suffused the Bay Area microcomputing hobbyist scene. See Isaacson, *Innovators*, 263–382; Johns, *Piracy*, 463–96; Levy, *Hackers*, 153–278; Turner, *From Counterculture to Cyberculture*.
53. Elizabeth Petrick similarly examines the extent to which commercial enterprise was embedded in the culture of the early Homebrew Computer Club specifically. Petrick, "Imagining the Personal Computer," 36–37.
54. Ceruzzi, *Modern Computing*, 240.
55. Some accounts of Ed Roberts's business dealings and the technical development of the Altair can be found in Freiberger and Swaine, *Fire in the Valley*, 31–35; Levy, *Hackers*, 186–92.
56. MITS, advertisement.
57. In some sources, the terms *microcomputer* and *microprocessor* were more or less synonymous—likely because there wasn't much to these machines at first *except* their microprocessor. For examples see the first issue of *BYTE* (September 1975), in which the article "Which Microprocessor For You" surveys popular microcomputer kits, the Altair included (described therein as "a package in which you can purchase an 8080 based system"). Similarly, the May 1975 issue of *Scientific American* shows a picture of a Teledyne Systems Company TDY 52 microprocessor on its cover but identifies that chip itself as a microcomputer rather than a microprocessor. Minicomputers might also be described as microcomputers if they were using microprocessors, as can be seen in the microprocessor manual for DEC's PDP-11 family of computing systems, advertised as "a 16-bit microcomputer with the speed and instruction set of a minicomputer." See Digital Equipment Corporation, *Microcomputer Handbook*, xv.

58. MITS, *Altair System*, 2.
59. Campbell-Kelly, *Airline Reservations*, 97–99. Software as a marketing service was especially common. The quality of software packages, especially for specific industries like banking or insurance, was a contributing factor in a business's decision to choose one mainframe company over another.
60. The clearest historical case study for this transformation from contract programming to software products is documented by Campbell-Kelly in his analysis of Applied Data Research's flowchart program *Autoflow*, which had originally been contracted by RCA in 1964. Campbell-Kelly, *Airline Reservations*, 100–101.
61. These prices are based on the documented cost of Informatics' Mark IV and Applied Data Research's *Autoflow*, both released in the 1960s. Campbell-Kelly, *Airline Reservations*, 106, 100. During this same period, in 1969, IBM agreed to unbundle the services it provided, software included, from the overall costs it charged its clients (a move undoubtedly bound to the US Department of Justice investigating IBM for antitrust violations, given that they commanded 70 percent of the US computing market). Campbell-Kelly, *Airline Reservations*, 109–10.
62. "Software Prices," 3. Loading the BASIC interpreter into the Altair's memory allowed a programmer to write code in BASIC that the Altair could process. An interpreter is essentially a translation of instructions for a programming language written in machine code, the lowest level of alphanumeric commands a computer can accept. As a specific implementation of a general programming language, an interpreter allows for the customization of the language to the memory limitations and microprocessor design of a specific type of computer—which was quite necessary in an era where memory was a precious commodity. In other words, in the 1970s there was no "one" BASIC: there were many BASICs, each customized for the hardware system it ran on. Syntax held across interpreters for the same language, but fine programming details often changed.
63. "Software Prices," 3. There was also a 4K BASIC bundle for $50. MITS, *Age of Altair*, 21.
64. Part of the consumer frustration over Altair's sale of BASIC was the fact that the programming language itself had been invented at Dartmouth and freely distributed to the broader computing community. For more information on the development of BASIC, see Rankin, *People's History*, 66–105.
65. As an Altair 8800–specific BASIC interpreter, TINY BASIC was customized for the Altair's Intel 8080 microprocessor and designed to run on only 2K of the Altair's minimum total of 4K memory (hence why it was "tiny"). Allison, Happy Lady, and Friends, "DESIGN NOTES," 15–18; Levy, *Hackers*, 231.
66. Allison, Happy Lady, and Friends, "DESIGN NOTES," 15.
67. A fairly detailed account of *Altair BASIC*, TINY BASIC, and *Altair BASIC* piracy can be found in Levy, *Hackers*, 225–29. In particular, Levy's account points out Dan Sokol as the source for duplicated paper tapes in the Bay Area and among the Homebrew Computer Club. For more details on the sharing ethos of homebrew hobbyist communities, see Driscoll, "Professional Work"; Johns, *Piracy*, chap. 16; Petrick, "Imagining the Personal Computer"; Freiberger and Swaine, *Fire in the Valley*, 140–43.
68. The 10 percent estimate is referenced in two sources: Allen, *Idea Man*, 93; Gates, "Open Letter," 2.
69. MITS's business model was set up to make money from hardware—not the Altairs themselves, which were nearly sold at cost, but the peripherals and expansion boards necessary to get the most out of the central hardware. Allen, *Idea Man*, 94.

70. *Altair BASIC*'s origin story is recounted in numerous texts, both academic and journalistic: Campbell-Kelly, *Airline Reservations*, 204–5; Ceruzzi, *Modern Computing*, 232–36; Cringely, *Accidental Empires*, 2–4; Levy, *Hackers*, 225–29.
71. Isaacson, *Innovators*, 332–33.
72. Manes and Andrews, *Gates*, 82.
73. Campbell-Kelly notes the uniqueness of this arrangement in *Airline Reservations*, 205.
74. Driscoll, "Professional Work," 263–64.
75. Driscoll, 264.
76. Driscoll, 265.

Chapter 2

1. For a survey of the Homebrew Computer Club's origins, demographics, and expressed values during the first two years of its existence, see Petrick, "Imagining the Personal Computer."
2. Wozniak, *iWoz*, 154.
3. The evening's activities were organized by Gordon French and Fred Moore. Moore, a consummate political activist, and French, an engineer with military-grade security clearance, forged their unlikely alliance out of shared grievance: both had been involved in educational initiatives at The People's Computer Company, and both felt the group's founder, computing evangelist Bob Albrecht, underappreciated their labor. Markoff, *What the Dormouse Said*, 273–76; Moritz, *Little Kingdom*, 103–4.
4. Wozniak, *iWoz*, 154. The full account of Wozniak's evening at the first Homebrew meeting, and future meetings, can be found in *iWoz*, chap. 10.
5. Wozniak, 155.
6. Moritz, *Little Kingdom*, 21.
7. Moritz, 29.
8. Wozniak, *iWoz*, 14.
9. Wozniak, 16.
10. Wozniak would become, he believes, the youngest person to ever hold a ham radio operator's license.
11. Malone, *Infinite Loop*, 16.
12. Barak, "John McCollum."
13. Moritz, *Little Kingdom*, 47–49.
14. Moritz, 46; Wozniak, *iWoz*, 98.
15. Where are the women?, you might wonder. Good question. Women were always there, somewhere, cast into the background of the stories men tell about themselves: permissive and pleasant mothers, wives frustrated and supportive by turns, sisters who plugged chips into circuit board sockets for a buck apiece, or women never named at all.
16. Moritz, *Little Kingdom*, 47–50.
17. Wozniak, *iWoz*, 119; Moritz, *Little Kingdom*, 119.
18. Wozniak, 56–118.
19. Wozniak, 75–92.
20. For accounts of the development of the blue box, see Moritz, *Little Kingdom*, 70–79; Isaacson, *Steve Jobs*, 27–30; Wozniak, *iWoz*, 93–118; Butcher, *Accidental Millionaire*, 25–33.
21. Moritz, *Little Kingdom*, 76; Wozniak, *iWoz*, 115.

22. Prices are given in Isaacson, *Steve Jobs*, 29. Jobs and Wozniak were "convinced that few women would be interested in their little device" (Moritz, *Little Kingdom*, 75). The pair's blue box escapades petered out after about a year. Jobs left the project first, "through a combination of boredom and fear of the possible consequences," leaving Wozniak to uncharacteristically run the business himself for a while (Moritz, *Little Kingdom*, 77). Moritz (*Little Kingdom*, 77–78) suggests their escapades lasted roughly a year, implying their efforts likely ended sometime in summer 1972—around when Jobs was about to leave for Reed College and Wozniak had become employed at Electroglass.
23. Moritz, *Little Kingdom*, 95. Tellingly, both Bill Fernandez and Wozniak were surprised to learn that Jobs had gotten himself a job at Atari because neither considered him much of an engineer.
24. Moritz, 95.
25. Accounts differ as to how Jobs got this assignment. According to Marty Goldberg and Curt Vendel, Nolan Bushnell, head of Atari, opened the opportunity to all Atari's engineers, but Jobs was the only person interested (they claim ball-and-paddle games were considered passé by the engineers) (*Atari Inc.*, 162–65). According to Walter Isaacson, Jobs was called into Bushnell's office and asked to design the game (*Steve Jobs*, 52). While it seems unusual that Bushnell would have interacted with Jobs, who was only a technician, Atari was still a small company at the time, and Al Alcorn, the Atari chief engineer who hired Jobs, has insisted that he was not even aware Jobs had been put on the assignment. (This is borne out in Goldberg and Vendel, *Atari Inc.*, 163; Alcorn, interview, 26).
26. According to Goldberg and Vendel, both Bushnell and Atari engineer Steve Bristow knew that giving the gig to Jobs meant Wozniak would complete it (*Atari Inc.*, 163). Some accounts even claim that Bushnell knew Wozniak was a great engineer because he had seen the home *Pong* clone Wozniak had designed with just twenty chips.
27. This estimate of how long it generally took to design the board for a new arcade game comes from Al Alcorn's recollections in Goldberg and Vendel, *Atari Inc.*, 163.
28. Goldberg and Vendel, *Atari Inc.*, 162.
29. Goldberg and Vendel, 165.
30. Moritz, *Little Kingdom*, 78.
31. Moritz, 100; Johnson and Smith, "Breakout Story."
32. Accounts differ as to how many chips were supposed to be saved. Isaacson claims that Bushnell would give a bonus for every chip under 50, but the evidence for this claim is unsourced (*Steve Jobs*, 52–53). Goldberg and Vendel report that the bonus was structured around every chip under 120. Their sourcing is largely based on direct interviews with key Atari personnel; see Goldberg and Vendel, *Atari Inc.*, 162. All things considered, Goldberg and Vendel's reporting seems more likely, as Isaacson's account would suggest that Bushnell was handing out a $1,000 bonus for each chip saved (roughly $4,700 today), and most arcade machines' chip counts were over 100, not between 50 and 100. For estimates of the bonus amount, Lee Butcher's *Accidental Millionaire* cites Wozniak claiming the number was $7,000. In *iWoz*, Wozniak claims it was "a few thousand dollars" (148).
33. Isaacson, *Steve Jobs*, 54.
34. Moritz, *Little Kingdom*, 36–40.
35. Isaacson, *Steve Jobs*, 31–52; Moritz, *Little Kingdom*, 86–101.
36. Hertzfeld, "Reality Distortion Field."

37. Documented instances of Jobs not upholding his bargains include his original deal with Paul Terrell for fully assembled Apple 1 computers. Terrell only received fully assembled printed circuit boards, which he took anyway (Moritz, *Little Kingdom*, 145). Stan Veit, a prominent New York City computer retailer, recalls accepting an offer from Jobs to receive an Apple 1 for consideration and getting stuck with a $500 cash-on-delivery charge (Veit, *History*, 89).
38. Wozniak, *iWoz*, 155.
39. Wozniak, 156.
40. There was another reason TV terminals were on Wozniak's mind over the summer of 1975: he had agreed to design one for Alex Kamradt, owner of the Mountain View–based time-sharing company Call Computer. Kamradt was an early participant at Homebrew; in the July 5, 1975, newsletter, he is acknowledged for providing the club with its own time-sharing account, as well as "giving us a good demo of his TVT [TV terminal]." According to Michael Moritz, Kamradt "wanted to rent or sell his customers a more convenient terminal with a typewriter keyboard that could be connected to a television." He hired Wozniak, and together they formed a subsidiary of Call Computer named Computer Conversor. Kamradt provided $12,000 in startup funding, while Wozniak promised to develop a terminal in exchange for 30 percent of the company and a free account at Call Computer. In Moritz's telling, Kamradt "saw the terminal as part of a grander scheme," as the springboard to developing a stand-alone computer. See Moritz, *Little Kingdom*, 116–26. While it is unclear exactly when Wozniak and Kamradt went into business together, schematics for the "Call Computer Terminal," held at Stanford University Archives, are dated June 22, 1975—just a few days prior to when Wozniak claims he had completed a fully functional microcomputer (see *iWoz*, 166). Like the TV terminal, the Computer Conversor is a vestigial tail of our media and computer history. Its near-simultaneous creation with Wozniak's first Apple 1 prototype likely explains why the Computer Conversor has been exiled from popular accounts: not only does it complicate the chronology, but it also decenters Steve Jobs from serving as the linchpin for the commercialization of Wozniak's homebrew computer.
41. Kirschenbaum, *Track Changes*, 124. In a professional context, the advantage of using a glass terminal was the lack of noise and the ability to do real-time text editing (a common feature for dedicated word processing installations).
42. Wozniak, *iWoz*, 153. Variations on such devices circulated in hobbyist magazines for several years, with the earliest documented example dating to the September 1973 issue of *Radio-Electronics*. Called the "TV Typewriter," the device was promoted as a "computer terminal for time-sharing services, schools, and experimental uses." Lancaster, "TV Typewriter," 43.
43. These features are didactically illustrated in Apple's first advertisement from 1976, where the "keyboard interface" and "complete video terminal electronics" connections are specifically labeled (see p. 60).
44. For a brief account of the 6502 and its inventor, Chuck Peddle, see Coughlin, "Chuck Peddle."
45. Wozniak, *iWoz*, 167.
46. Wozniak recounts his experience demonstrating the Apple 1 to Homebrew in *iWoz*, 184–85. While the Apple 1 could support up to 8K of RAM, it was sold with only 4K.
47. While several historical accounts of the Apple 1 claim that it provided sixteen times the memory of an out-of-the-box Altair, this claim is somewhat disingenuous. It is true that the bare-bones Altair kit contained only 256 bytes of memory, but nearly

everyone had to upgrade in order for the machine to be useful. 4K memory was considered the smallest amount to make the machine useful—especially given that 4K memory was the minimum requirement to run Micro-Soft's *Altair BASIC*.

48. Moritz, *Little Kingdom*, 127.

49. "Member Microcomputer Systems."

50. Moritz, *Little Kingdom*, 136.

51. Moritz, 137.

52. Mortiz, 135–36.

53. According to Moritz, Jobs also had to convince Wozniak to give up his deal with Call Computer. *Little Kingdom*, 140.

54. The original partnership was actually between three people: Wozniak, Jobs, and Ron Wayne, whom Jobs knew through Atari. Wayne designed Apple's original illustrative logo but was officially responsible for mechanical engineering and documentation, in exchange for a 10 percent stake in Apple Computer. Wayne left the partnership after less than two weeks, selling his share to Wozniak and Jobs for $800. Moritz, *Little Kingdom*, 139–40; Wozniak, *iWoz*, 174, 185.

55. Wozniak's autobiography claims that the presentation to the Homebrew Computer Club happened in March 1976 and that it happened after Terrell made them a deal. This contradicts Isaacson. Wozniak, *iWoz*, 177, 184; Isaacson, *Steve Jobs*, 66–67. Beyond the knowledge that Terrell was a frequent attendee at Homebrew and that he made a deal with Wozniak and Jobs for a hundred Apple 1s, the exact chain of events is left to future scholars to decipher.

56. Moritz, *Little Kingdom*, 142–43.

57. Moritz, 150–52.

58. For Wozniak's account of developing BASIC for the Apple 1, see *iWoz*, 180–84. Knowing little about how to develop language interpreters, Wozniak based his implementation of BASIC on HP BASIC (rather than *Altair BASIC*) and also used integers instead of floating point arithmetic. This BASIC was later updated and implemented on the Apple II; however, the lack of floating point was problematic for some programmers. Apple later tried to correct this with the release of Applesoft BASIC. For more information, see Weyhrich, *Sophistication*, 47–53.

59. Moritz, *Little Kingdom*, 153.

60. Wozniak, *iWoz*, 188.

61. Wozniak, 189–90.

62. Accounts differ with regard to how many Apple 1s sold at PC '76. Stan Veit, who shared a booth with Wozniak and Jobs, claims none were sold. Wayne Green, who was also at the fair but at a separate booth and was given his figures by Jobs, claims twenty Apple 1s were sold. Veit, *History*, 97; Green, "Remarks from the Publisher," 6.

63. Wozniak mentions the Atari pitch in *iWoz*, 196; Moritz describes the Commodore deal in *Little Kingdom*, 161–62.

64. Moritz, *Little Kingdom*, 174.

65. Nicholas, *VC*, 187.

66. For a summary of Valentine's investment style and how he fits into the history of American venture capital, see Nicholas, *VC*, 222–31.

67. Moritz, *Little Kingdom*, 175.

68. Moritz, 176.

69. Moritz, 177.

70. Isaacson, *Steve Jobs*, 79–80.

71. The first announcement of the West Coast Computer Faire can be found on the front page of the September 1976 issue of the Homebrew Computer Club newsletter.
72. Ahl, "First West Coast Computer Faire," 24.
73. The location of Apple's booth can be discerned by cross-referencing the exhibitor list with the floor map in the event program; see Computer Faire, program, 36, 42.
74. Moritz, *Little Kingdom*, 188, 191.
75. Moritz, 191–92; Veit, *History*, 96–97.
76. Moritz, *Little Kingdom*, 186–91.
77. Stein, "Domesticity," 196.
78. Moritz, *Little Kingdom*, 188.
79. Wozniak's choice was prescient: the 6502 would eventually make its way into a host of popular consumer appliances, including the Atari VCS and the original Nintendo.
80. For information on early BASIC, see Weyhrich, *Sophistication*, 47–51.
81. A cassette deck was not included as part of the microcomputer purchase; Wozniak worked on the assumption hobbyists would provide their own decks. Weyhrich, *Sophistication*, 42–43.
82. Games were always a core interest for Wozniak; one of his earliest programming exercises for the Apple II prototype was to create a version of Atari's *Breakout*. Wozniak, *iWoz*, 190–92.
83. Moritz, *Little Kingdom*, 193.
84. RadioShack did release a Model I expansion interface in 1978, which allowed users to add RAM and peripherals.

Chapter 3

1. Listing drawn from Mad Hatter Software advertisement, *BYTE*, March 1979, 163.
2. Rosen, "VisiCalc," 1.
3. Rosen, 2.
4. Rosen, 2.
5. The "killer app" hypothesis, as Martin Campbell-Kelly explains, argues that "a novel application, by enabling an activity that was previously impossible or too expensive, causes a new technology to become widely adopted." *Airline Reservations*, 212. Many popular press and retro enthusiast writings on *VisiCalc* maintain this mythological stance, claiming the software is the reason we have a personal computing industry. Such claims, of course, are overwrought. As Campbell-Kelly writes, "The personal computer revolution would have happened with or without *VisiCalc*. However, it is plausible that *VisiCalc* accelerated the process by several months." For further dissection of the "killer app" hypothesis, see Campbell-Kelly, *Airline Reservations*, 212–14.
6. Carroll, *It Seemed Like Nothing Happened*, 207–32.
7. Carter, "Energy and the National Goals."
8. Harvey, *Neoliberalism*, 2.
9. The tech journalist Steven Levy references this as a primary observation from Allerton Cushman's pamphlet, "Confessions of an Apple Byter." Levy, "Spreadsheet Way of Knowledge."
10. Levering, Katz, and Moskowitz, *Computer Entrepreneurs*, 130.
11. For more information on Bricklin's background, see Bricklin, *Bricklin on Technology*, 423–26; Levering, Katz, and Moskowitz, *Computer Entrepreneurs*, 129–31.
12. Campbell-Kelly, "Number Crunching," 7.

13. For references to these various anecdotes, see Bricklin, *Bricklin on Technology*, 424, 426; Campbell-Kelly, "Number Crunching," 7; Cringely, *Accidental Empires*, 64–66; Grad, "Creation," 21; Levering, Katz, and Moskowitz, *Computer Entrepreneurs*, 131.
14. The use of computing in business settings traditionally starts with the 1951 UNIVAC. Initially conceived to aid the automation of census calculations for the US government, the system was soon thereafter bought by a market research firm, an insurance company, and other corporate operations. Campbell-Kelly et al., *Computer*, 99–103.
15. For an overview of the midcentury software products industry in the United States, see Campbell-Kelly, *Airline Reservations*, chaps. 4 and 5.
16. Roughly two-thirds of that revenue derived from industry-specific and cross-industry applications. Campbell-Kelly, *Airline Reservations*, 126.
17. Campbell-Kelly, 126–27.
18. Grad, "Creation," 21.
19. Burt Grad provides a deeper overview of the influence the Typeset-10 and the Harris 2200 had on Bricklin's design ethos. "Creation," 21–22.
20. Campbell-Kelly, "Number Crunching," 7. While most sources claim Bricklin and Fylstra were introduced by an HBS professor, Paul Freiberger and Michael Swaine claim Frankston already knew Fylstra, having converted a bridge program to the Apple II for Personal Software (*Fire in the Valley*, 229).
21. Fylstra, "Personal Account," 4.
22. Fylstra, "User's Report"; Fylstra, "Radio Shack TRS-80."
23. Dan Bricklin and Bob Frankston, Interview by Martin Campbell-Kelly and Paul Ceruzzi, May 7, 2004.
24. For a summary of *VisiCalc*'s design and implementation, see Grad, "Creation," 24–26.
25. Bricklin and Frankston, interview, 15; Grad, "Creation," 23; Bricklin, "Special Short Paper."
26. Both Frankston and Fylstra have taken credit for coming up with the name "*VisiCalc*." See Bricklin and Frankston, interview, 40; Fylstra, "Personal Account," 8.
27. Bricklin discusses his development process on the Apple II in *Bricklin on Technology*, 427.
28. Accurate sales figures are difficult to determine. These estimates are taken from Weyhrich, *Sophistication*, 55.
29. Fylstra, "Personal Account," 5. Personal Software's sales data also informed Fylstra that the emerging consumer computer franchise Computerland (which carried Apple products) was gaining ground compared to nonfranchised stores.
30. Grad, "Creation," 26. *VisiCalc*'s high memory requirements were due to its mode of simultaneous calculation, as well as screen scrolling, which permitted users to navigate a workspace larger than their monitor; such information needed to be held in memory, even if it wasn't viewable to the user.
31. Weyhrich, *Sophistication*, 29.
32. For an overview of Apple's floppy disk mass storage peripheral, the Disk II, see Weyhrich, *Sophistication*, 55–61.
33. Helmers, "Magnetic Recording Technology," 6–8.
34. Floppy disks were available for the Altair but were largely regarded as business products. MITS's 1975 Christmas catalog priced a disk drive and controller at $1,980 assembled, a price far outside what was reasonable for general hobbyists. MITS, "MITS-mas," 6.

35. For primary documents discussing cassette-based storage media of the late 1970s, see Willis, Smithy, and Hyndman, *Peanut Butter*, 59–64; Electronic Data Systems Corporation, *Little Computers*, 48–50; Freiberger and Chew, *Consumer's Guide*, 31–32.

36. For a theoretical analysis of magnetic data storage, see Kirschenbaum, *Mechanisms*, chap. 1. For explicit discussion of the concept of inscription, see 58–68.

37. The Commodore PET's designers even went so far as to install the cassette player into the chassis of the computer (good for leaning into the idea of an all-in-one consumer purchase but inevitably ensuring the deck couldn't be upgraded).

38. Weyhrich, *Sophistication*, 42.

39. Willis, Smithy, and Hyndman, *Peanut Butter*, 65.

40. Electronic Data Systems Corporation, *Little Computers*, 49.

41. According to some reporting, *VisiCalc* did make accommodations for storing user data on cassette tape, but these features were quickly dropped. "VisiCalc: User-Defined Problem Solving Package."

42. For information on the development of the floppy disk, see Weyhrich, *Sophistication*, 56; Ceruzzi, *Modern Computing*, 231–32; Bates, "Floppy Disk," F11. For an address of the industrial life span of the floppy disk, see Amankwah-Amoah, "Competing Technologies."

43. For a contemporaneous explanation of how floppy disks work, see Shelly and Cashman, *Computer Fundamentals*, 9.1–9.6.

44. Hoeppner, "Interface," 72.

45. Weyhrich, *Sophistication*, 60.

46. Apple's 5.25-inch disk drives were released as the Disk II in July 1978, with a capacity of 113K—20K more than Shugart's drives offered standard. This enhancement in storage volume was due to the custom hardware and software Wozniak oversaw for the Disk II. For documentation of the TRS-80's expansion interface unit, see Radio Shack, *Expansion Interface*.

47. According to Weyhrich, the Disk II had an introductory price of $495, including the controller card, if users bought it before Apple had the drives in stock. Once the Disk II was released, the price was $595. Weyhrich, *Sophistication*, 60. Wozniak also did extensive customization to the controller software that managed the drive's internal processes, resulting in faster-than-normal floppy disk access times. See *iWoz*, 212–17.

48. Rose, *West of Eden*, 62.

49. For editorial coverage of RadioShack's restrictions on selling third-party software, see Green, "Painful Facts."

50. Examples of this analogy can be found in Campbell-Kelly, "Number Crunching," 7; Carlston, *Software People*, 70.

51. Hobbyist duplication practices and software piracy existed as gray markets alongside retail software. See chapter 5 for more on this subject.

52. For information on Personal Software's founding, see Campbell-Kelly, "Number Crunching," 7; Tommervik, "Exec Personal," 6.

53. See the Personal Software ad in *BYTE*, March 1978, 118.

54. See ads in *BYTE*, March 1978, 118; and April 1978, 170.

55. Fylstra, "Personal Account," 7.

56. It was not uncommon for royalty percentages to be slightly different on wholesale revenue and direct sales, as publishers made slightly different margins on the two revenue flows. The majority of a publisher's software revenue was typically wholesale, which was usually set at 50 to 70 percent the retail price; wholesale royalties were paid on the revenue generated by selling software to the distributor, not the retail price. However, at this time, the industry was small enough that most publishers also sold direct to consumers, typically via mail order. When selling direct, they sold at retail price, so as not to create a competitive price difference with brick-and-mortar retailers. This meant that the publisher absorbed all the markups traditionally afforded to the distributor and retailer in the supply chain. In practice, this meant a publisher made more revenue per unit selling their software direct to consumers—though these gains were offset by the cost of workers needed to fulfill orders and the economies of scale necessary to create a low manufacturing price point (which was possible only through the wholesale revenue that distributors created through bulk purchasing). For an example of a distributor price list from this period, see Brøderbund Dealer Price List, February 1, 1982. Strong Museum of Play, Brøderbund Software, Inc. Collection, Subseries E: Sales and Marketing, Box 8, Folder 2.

57. Many early publishing outfits were so small that they were simply composed of the software programmed by a single individual, such as Don Alan Ltd. or Robert C. Kelly. Others included Ed-Pro, Digital Research, Software Records, AJA Software, National Corporate Sciences, Structured Systems Group, Sunshine Computer Company, and Contractor's Management Systems. All examples can be found in *BYTE*, May 1978.
58. For more information on the development of *Microchess*, see Freiberger and Swaine, *Fire in the Valley*, 134; Jennings, interview, 10–19.
59. For "Gold Cassette" reference, see ad in *BYTE*, June 1979, 142.
60. Tommervik, "Exec Personal," 7.
61. Fylstra, "Personal Account," 6.
62. Fylstra.
63. According to Jennings, the rate was informed by the royalties he himself had on *Microchess*. See Jennings, "*VisiCalc*—The Early Days."
64. Grad, "Creation," 27.
65. Grad.
66. Tommervik, "Exec Personal," 7.
67. The time-sharing system was a Prime 350 minicomputer, which had a PL/1 compiler similar to the time-sharing system Bricklin and Frankston had used at MIT. It was not unheard of to develop microcomputer software in a custom minicomputer environment, which allowed for more deployment of programming utilities and tools. See Bricklin, *Bricklin on Technology*, 447; Grad, "Creation," 26; Jennings, "*VisiCalc*—The Early Days."
68. See Jennings, "*VisiCalc* 1979," 1, 2. There is no archival evidence that Personal Software conducted any user testing beyond the development team (though apparently Frankston and Jennings both used *VisiCalc* to compile their 1978 taxes).
69. See ad in *BYTE*, September 1979, 50.
70. Fylstra, "Personal Account," 9–10.
71. Veit, *History*, 99.
72. Veit, 100.
73. Bricklin, *Bricklin on Technology*, 452. This quote was originally transcribed by Bricklin from a speech given by Adam Osborne when awarding *VisiCalc* the 1979 White Elephant Award for people who changed the industry.
74. Fylstra, "Personal Account," 9. "Whole product" was a concept developed by Regis McKenna. For more information on Regis McKenna's marketing strategies, see McKenna, *Regis Touch*.

75. According to Bricklin, the manual was drafted three times: first by Bricklin, then by a freelancer, and finally by Fylstra. Personal Software handled everything except the printing of the reference card; that was reproduced by Bricklin's father, Baruch Bricklin, who owned a print shop in Philadelphia.
76. Fylstra, "Personal Account," 10.
77. This image can be found in *BYTE*, September 1979.
78. Jennings, "*VisiCalc* 1979," 3.
79. For more on the development of the mainframe and minicomputer software industry, see Campbell-Kelly, *Airline Reservations*, chap. 5.
80. For further explanation of the structure of data processing departments and the challenges microcomputers posed to those employees, see Beeler, "Personal Computing."
81. Helmers, "Tower of Babel," 156.
82. McMullen and McMullen, "Screen Envy," 277–78.
83. McMullen and McMullen, 278.
84. McWilliams, *Personal Computer Book*, 60.
85. Apple was so esteemed as an up-and-coming entrepreneurial company, it was featured in the 1979 premiere issue of *Inc.* See Sklarewitz, "Born to Grow."
86. "Softalk Presents the Bestsellers," *Softalk*, January 1984. This edition of *Softalk*'s best-seller list reported that *VisiCalc* slipped from number four (its previous month's ranking) to number eleven. *Softalk* attributed the slippage to several factors, including the strong showing of entertainment software in the holiday season, competition from other products such as *Multiplan*, and what *Softalk* identified as the "changing profile of new owners," who were "more directed toward word processing than toward number crunching" (267).
87. "Softalk Presents the Bestsellers," *Softalk*, December 1981, 205.
88. See Sigel and Giglio, *Guide to Software Publishing*, 49–50. While Sigel and Giglio document, in table 3.11, that the *value* of consumer purchases for the year was $905 million, publishers would not have captured more than 50 percent due to the prevalence of discounts and OEM bundling (placing publisher revenues closer to half a billion dollars).
89. Sigel and Giglio, 47.
90. Sigel and Giglio, 46; emphasis in original.
91. Personal Software, "*VisiCalc*," 2.
92. Levy, "Spreadsheet Way of Knowledge."
93. Levy.
94. Levy.
95. Levy.
96. Deringer, "Michael," 56.
97. Harvey, *Neoliberalism*, 2.
98. Allen, "Official Response," 42.
99. Dan Fylstra, quoted in "Tradetalk," *Softalk*, March 1982, 43.
100. For sources on the lawsuit between VisiCorp and Software Arts, see Pollack, "Software Winner"; Caruso, "VisiCorp, Software Arts."
101. Pollack, "Software Winner."
102. Kapor had developed VisiCorp's complementary *VisiCalc* products, *VisiPlot* and *VisiTrend*. When Kapor decided to leave VisiCorp, he was bought out of his royalties on these products for $1.2 million—a price VisiCorp hastily paid, and without challenging Kapor's written addendum to his buyout agreement, which permitted

him to work on his own spreadsheet product. More information on Kapor and *Lotus 1-2-3* can be found in Levering, Katz, and Moskowitz, *Computer Entrepreneurs*, 188–95; Campbell-Kelly, *Airline Reservations*, 216; Hixson, "Lotus 1-2-3."

103. Caruso, "Software Gambles."
104. Pollack, "Visicorp Is Merging into Paladin."

Chapter 4

1. Exceptions necessarily prove the rule. As Rankin has documented, many students in Minnesota had in-school access to software for educational and exploratory purposes, games included, via a statewide time-sharing network. Yet the point holds that such networks were never exclusively designed for or dedicated to game-playing. See Rankin, *People's History*, chap. 5.
2. As the computing historian Michael R. Williams has said, "There is no such thing as 'first' in any activity associated with human invention. If you add enough adjectives to a description you can always claim your favorite." Quoted in Arthur, *Nature of Technology*, 126. Yet Roberta Williams is undoubtedly part of a very small cast of women developers who created commercial computer or console games during the late 1970s and early 1980s. Other notable women in this category include Carol Shaw, Dona Bailey, and Joyce Weisbecker.
3. An extensive account of the making of *Mystery House* can be found in Nooney, "Let's Begin Again."
4. Levering, Katz, and Moskowitz, *Computer Entrepreneurs*, 236.
5. Levy, *Hackers*, 295; Nooney, "Let's Begin Again," 79-80; Roberta Williams, in-person interview with author, January 14–15, 2019.
6. Levy, *Hackers*, 284–87.
7. Nooney, "Let's Begin Again," 80. While some sources acknowledge Roberta had done a little COBOL programming, her exposure to computing was more extensive than has been documented previously. Roberta Williams, in-person interview with author, January 14–15, 2019.
8. For a more extensive discussion of *ADVENT*'s creation and how the Williamses accessed the program, see Nooney, "Let's Begin Again," 74–83.
9. Intriguingly, descriptions of *ADVENT* from the DEC user group catalog described the program as "more of a puzzle than a game" owing to its lack of replayability. See Digital Equipment Computer Users Society, *DECUS Program Library*, 47.
10. Levy, *Hackers*, 296.
11. *Cathexis* is a term taken from David Sudnow's *Pilgrim in the Microworld*. Other examples of players' fixation on games can be found in Brand, "SPACEWAR."
12. Albert et al., interview, 9; Levy, *Hackers*, 294; Tommervik, "On-Line Exec," 4.
13. Though the reason for the couple's choice of the Apple II is not documented, they were likely swayed by the same features that led Dan Fylstra and Dan Bricklin to the platform. If you were a serious programmer looking for a microcomputer in late 1979 or early 1980, the Apple II was the obvious choice because of its robust technical capabilities, openness to independent developers, early to market floppy disk peripheral, and relatively inexpensive price.
14. For deeper documentation of what microcomputers the Williamses had exposure to, see Nooney, "Let's Begin Again," 83–84, esp. notes 26 and 28.
15. Levering, Katz, and Moskowitz, *Computer Entrepreneurs*, 240.
16. Reference to *Clue* and/or *And Then There Were None* (sometimes referred to as *Ten*

Little Indians) frequently arises in primary documents from the period: see Levering, Katz, and Moskowitz, *Computer Entrepreneurs*, 240; Levy, *Hackers*, 297; Tommervik, "On-Line Exec," 5. However, it is unclear what specific role either the game or the novel played in Roberta's design process.

17. As Wozniak put it, "The Apple II was also an ideal computer for anybody who wanted to design a computer game." *iWoz*, 209.
18. As discussed in chapter 2, Wozniak had a stint helping Jobs in his work at Atari and also built his own computer system exclusively to play a personally programmed version of *Pong*. Wozniak, *iWoz*, 141–47.
19. North, "Apple II Computer," 28. Wozniak even designed the system's first game, a *Breakout* clone variously known as *Brickout*, *Breakthrough*, or *Little Brickout*, sold on cassette with the first Apple IIs and later included as a demo with the DOS 3.2 system update, along with a variety of other games and interactive programs.
20. The Apple II did not have a dedicated color graphics chip. Rather, Wozniak designed the Apple II to produce color by manipulating the color bursts of black-and-white signals, which involved "exploiting quirks in how TVs decode analog video signals," as Stephen Cass put it in an interview with Al Alcorn. Alcorn, the engineer who built Atari's *Pong* machine, claims Wozniak learned this technique from him. For more information, see Cass, "Al Alcorn."
21. Instead of expanding these systems into color, these manufacturers released entirely new, incompatible systems, the Tandy Color Computer and the Commodore VIC-20, in 1980.
22. The Apple II's low-resolution mode was still competitive with, if not better than, the resolutions offered by its competitors, the PET (25 × 40) and the TRS-80 (16 × 64). See Wallace, "PET vs. the TRS-80." Both of the Apple II's resolution modes technically had room for four lines of text at the bottom of the screen, which could be traded off for additional graphical space. Additionally, while the low-res mode technically provided sixteen colors, two of the grays were indistinguishable on CRT screens of the era, dropping usable colors to fifteen. See Finnigan, *New Apple II*, 294.
23. Finnigan, *New Apple II*, 307.
24. The original BASIC interpreter for the Apple II, known as Integer BASIC, lacked support for floating-point mathematics, limiting its use for a variety of applications. Part of the release for the Apple II Plus was to offer an onboard BASIC with floating point support, which was called Applesoft. For more information on Apple's BASICs, and the Apple II Plus, see Weyhrich, *Sophistication*, 47–53, 115–23. For a comparison of the Apple II and the Apple II Plus, especially from a programmer's perspective, see Carpenter, "Apple II vs Apple II Plus."
25. Weyhrich, *Sophistication*, 116.
26. This approach had precedent: hi-res mode made use of shape tables, which allowed users to replicate standardized shapes using vector coordinates.
27. If a player runs the game on original hardware or an appropriate emulator, the lines will generate on screen in the order Roberta Williams drew them. See Nooney, "Let's Begin Again," 86.
28. Ken Williams, phone interview with author, October 8, 2013.
29. For further reading on the arcade era in the United States and arcade cabinet design, see Guins, *Atari Design*; Kocurek, *Coin-Operated Americans*.
30. For further reading on the domestic dynamics of the American video game console industry, see Newman, *Atari Age*, esp. chaps. 3 and 4.

31. For more information on the economics of home consoles, see Williams, "Structure and Competition," 44.
32. RadioShack advertised the TRS-80 for $499, while an Atari VCS was regularly $164 and the Mattel Intellivision was $269 at JCPenney. RadioShack, *1980 Catalog*, 170; JCPenney, *Christmas 1980*, 556–57.
33. The game development studio Activision was the first to succeed in bringing third-party game development competition to the console market. For more information on the impact this had on Atari, see Donovan, *Replay*, 89–93. In order to manage third-party development, the console industry later converted to licensing arrangements. See Williams, "Structure and Competition," 44–45.
34. As discussed in chapter 3, competitors like RadioShack discouraged third-party software development by restricting the sale of such software in RadioShack stores. Thus, while the TRS-80 had a large software catalog by 1980, RadioShack's closed distribution meant none of it could be purchased at RadioShack stores without RadioShack's permission (and financial cut).
35. *VisiCalc* was an exception: part of the promotional effort for that product was to give buyers a sense of confidence, so Fylstra and Jennings advertised it in mainstream business magazines. This was, however, exceedingly uncommon in the early days of microcomputer software development.
36. The MOS 6502 was the microprocessor used on the Apple II as well as the Commodore PET and a variety of more obscure trainer systems such as the SYM-1 and the AIM-65. Ads for all these systems can be found in the same May 1980 issue of *MICRO* the Williamses advertised in.
37. Levering, Katz, and Moskowitz, *Computer Entrepreneurs*, 240.
38. Levy, *Hackers*, 299.
39. Levy. A review and a rare screenshot of *Skeetshoot* (referred to as *Skeet*) can be found in Lubar, "Software," 24.
40. For theorization of the screenshot in video game history, see Gaboury, "Screen Selfies."
41. Levy, *Hackers*, 299.
42. Levy, 300.
43. Albert et al., interview, 19; Levy, *Hackers*, 300.
44. Levy, *Hackers*, 300.
45. Albert et al., interview, 19.
46. Ken sold the business to Bob Leff, who used the inventory and accounts receivable to found Softsel, which quickly became the largest computer software distributor in America. Tommervik, "Exec Softsel," 4–6, 47–48.
47. Nooney, "Let's Begin Again," 87.
48. Lubar, "Software," 24.
49. "Softalk Presents the Bestsellers," *Softalk*, October 1980, 27. Styled after *Billboard*'s "Hot 100" weekly pop charts, *Softalk*'s best-sellers feature offered a quantitative, month-to-month ranking of the top thirty best-selling units of Apple II software, accompanied by an index number registering the "relative marketplace strength" of each ranked product. While reader surveys were not uncommon in microcomputer enthusiast magazines, no magazine had yet attempted to track software sales performance or market strength. As a routine snapshot of consumer spending, *Softalk*'s best sellers list gave software publishers information they couldn't acquire from their own sales records alone—a sense of both how companies performed comparatively and what types of software was actually selling.

50. The remaining eight pieces of software represented three utilities, two word processors, two data management/small business packages, and one piece of typing instructional software.
51. Jaroslav Švelch makes a similar observation about games as consumable software in Švelch, *Gaming the Iron Curtain*, 138.
52. International Resource Development, *Microcomputer Software Packages*, 14, 159. Sigel and Giglio concur in estimating the 1983 software market at $1.5 billion in customer spend; see Sigel and Giglio, *Guide to Software Publishing*, 5.
53. International Resource Development, *Microcomputer Software Packages*, 159.
54. The few publishers that solely released games, such as Infocom and Adventure International, typically specialized in text adventures. It was much easier to port text-only games to other systems, meaning that, for these companies, diversification and expansion came through releasing to every system rather than moving into other software categories.
55. The term *borrowed* was used extensively in Eklund's interview with Albert et al., Smithsonian Institution Archives.
56. This kind of competition between identical games could exist because in the very early microcomputer industry, distribution was still fairly decentralized. While major chains like Computerland likely limited the number of identical games they stocked, some games might only be distributed regionally, in non-chain computer shops, or sold predominantly through mail order.
57. Several of On-Line's popular cloned Apple II games include *Threshold* (*Galaga*), *Gobbler* (*Pac-Man*), *Pegasus II* (*Scramble*), and *Cannonball Blitz* (*Donkey Kong*). These practices did attract heat to some publishers, including On-Line Systems, Brøderbund, and Stoneware. Atari threatened Ken Williams with litigation over On-Line's release of a *Pac-Man* clone called *Jawbreaker*. For more information, see Levy, *Hackers*, chap. 16; Tommervik, "Great Arcade/Computer Controversy."
58. Švelch draws a formal distinction between ports and conversions as programming practices, see Švelch, *Gaming the Iron Curtain*, 164–66. Porting was not unique to microcomputers; several sources trace the practice to mid-1970s minicomputing, though other precedents seem likely. My thanks to Matthew Hockenberry, Jaroslav Švelch, and Jose Zagal for pointing me to these sources on Twitter. See Malcolm, "Real-Time," 3–4; Tanenbaum, "Guidelines."
59. Advertisements also indicate Apple II publishers converted their products to the TRS-80 or the TI 99/4A, though to a lesser extent than MOS 6502–based machines. It is unclear whether these would have been ports, which made some use of the original code base or simply reprogrammed from the ground up.
60. For a thorough account of the arrival of venture capital to the entertainment software scene, see Carlston, *Software People*, chap. 8.
61. Metz, "Pac-Man and Beyond."
62. Investor interest was also fueled by changes in tax law, specifically the Revenue Act of 1978 and the Economic Recovery Tax Act of 1981, both of which effectively reduced taxes on long-term capital gains. See chap. 7.
63. The Harvard Business School professor Tom Nichols has made the acute observation that the distribution of returns in venture capital largely mirrors that of the early nineteenth-century whaling industry. In whaling, the vast majority of ships either failed to break even or never returned it all, while a select few brought back astronomical returns for their investors. See Nichols, *VC*, chap. 1.
64. Levy, *Hackers*, 359.

65. Campbell-Kelly, *Airline Reservations*, 225.
66. These tensions are perhaps most famously documented in Levy's embedded reporting on Sierra On-Line, see Levy, *Hackers*, pt. 3.
67. Quote from Edmund Auer, president of CBS software. Larson, "Many Firms," 23. Somewhat confusingly, CBS also launched CBS Video Games (perhaps better known as CBS Electronics) in 1982 as a subsidiary of their toy division. CBS Video Games focused on the console market, producing cartridges for the Atari 2600. Some of CBS's advertising suggests it imagined similar synergies between its video games division and its broader media reach, based on its ability to engage product placement both on television and in magazines. See CBS, "CBS Video Game Plan." CBS Video Games/CBS Electronics appears to have been shut down roughly around the time of the North American Video Game Crash in 1983.
68. Larson, "Many Firms," 23.
69. Carlston, *Software People*, 206.
70. While never released, a series of adventure games based on the *Family Circus* cartoons was also planned. Gear, "Backtalk," 122–24.
71. Rosenthal, "Atari Speaks Out," 58.
72. Sipe, "Computer Games," 10–11.
73. Nooney, "The Uncredited," 129.
74. Marie (Cavin) Iden, phone interview with author, September 3, 2015.
75. Lock, "Editor's Notes," 6.
76. "Showgirls," 269.
77. Statistics and industry analysis from Kneale, "Overloaded System," 1, 13.
78. For various treatments of the 1983 North American Video Game Crash, see Cohen, *Zap*; Donovan, *Replay*, chap. 8; Guins, *Game After*, 220–30; Kline, Dyer-Witheford, and De Peuter, *Digital Play*, 103–6.
79. Sigel and Giglio, *Guide to Software Publishing*, 40.
80. Campbell-Kelly, *Airline Reservations*, 226.
81. Crawford, *Art of Computer Game Design*, 54, 44.
82. Nelson, *Home Computer Revolution*, 154.

Chapter 5

1. The original ad appears in the January 1981 issue of *MICRO: The 6502 Journal*, 80.
2. Tripp, "Copyright," 5.
3. The granularity of such distinctions is illustrated in the catalogs of the DECUS Program Library, a software index from the 1960s and 1970s that served users of DEC's popular line of PDP minicomputers. "Utility" is just one of about twelve specialist categories of software related to minicomputing operations. See DECUS, *DECUS Program Library Catalog* (November 1969), ii; DECUS, *PDP-8 Program Library Catalog* (June 1979), 149.
4. Gates, "Software Contest," 1, 3. The software contest was a regular feature in the first volume of *Computer Notes* and the first half of the second volume. It's worth noting that *Computer Notes* disambiguated between "program" submissions and "subroutine" submissions and that utilities could be found in both.
5. *Altair Users Group Software Library*, i.
6. Hallen, "INDXA."
7. Hallen, 116.
8. I have amended Hallen's actual language to make this example clearer for contem-

porary readers. Hallen referred to his individual programs as "routines," which he aimed to group together "on a single tape as a continuous program." In Hallen's words, the INDXA utility works by creating a single file out of multiple routines. Hallen, 116.

9. Hallen, 119.
10. Southwestern Data Systems, "Apple-Doc." For more on Southwestern Data Systems and its founder, Roger Wagner, see Tommervik, "Assembling Useful Utilities," as well as Wagner's personal website, https://rogerwagner.com/.
11. *Bill Budge's 3-D Graphics System*, 1. In modern parlance, *Bill Budge's 3-D Graphics System* was a tool for creating and managing graphics libraries. Any games created with the software for resale were required to provide a credit notice on the packaging, in documentation, and in the game itself.
12. *Bill Budge's 3-D Graphics System*.
13. Yuen, "Head's Up!," 72.
14. For extensive documentation of the legal history of software in the United States, see Con Díaz, *Software Rights*; for discussion of the first software copyrights, see 63–64. For a review of software's legal status, as understood in the early 1980s, see Mihm, "Software Piracy."
15. Con Díaz, *Software Rights*, 163.
16. For more information on the use of copyright in the early microcomputer industry, see Con Díaz, 161–84.
17. This language is drawn from two different reproductions of *Microchess*'s KIM-1 documentation: a manual for a copy published by *The Computerist* and a manual for Jennings's self-published version (reproduced in raw text on Jennings's website). See Jennings, *MicroChess for the KIM-1*, 2; Jennings, *MICROCHESS*.
18. Jennings, *MicroChess for the KIM-1*, 2; Jennings, *MICROCHESS*.
19. Fylstra and Raskin, *Microchess*, 2.
20. For an exploration of the perceived risks domestic pirating posed, see Johns, *Piracy*, chap. 15.
21. For information on the Hacker Ethic, see Levy, *Hackers*, 39–51.
22. I'm indebted here to the work of retrocomputing enthusiast and programmer Andrew McFadden, who scrupulously documented the experience of negotiating cassette-based copy protection on his website; see McFadden, "Early Copy Protection."
23. For an overview of academic writing on copy protection (most of which is focused on game software), see Aycock, *Retrogame Archaeology*, chap. 7; O'Donnell, "Production Protection"; Hodges, "Technical Fixes"; Kelty, "Inventing Copyleft."
24. The Apple II floppy disk organization transitioned from 13 to 16 tracks with the DOS 3.3 update, which was released in August 1980. Worth and Lechner, *Beneath Apple DOS*, 2–3.
25. This common illustration is somewhat inaccurate. Tracks do not rub up against each other, and sectors are actually offset from one another. For illustrations of this, see Chamberlain, "Apple II Copy Protection."
26. While many sources detail that floppy disk sectors held 256 bytes, and I have replicated that description here for simplicity's sake, it is not entirely technically accurate to speak of the data held on a disk in terms of bytes. I have benefited in my understanding of floppy disk data organization and copy protection from in-depth email conversations with the software hacker 4am (https://twitter.com/a2_4am), who explained to me that the use of the term *bytes* conflates "bytes" (such as bytes in

memory; a sector does contain 256 such bytes) with what—in the Apple II world—are generally called "nibbles" (8-bit chunks of the bitstream that is actually stored on the physical disk). Due to limitations of the Disk II hardware, what starts as 256 bytes in memory is translated into 342 nibbles that are actually stored on disk, then translated back when loaded from disk. It is these 342 nibbles that are prefixed by the special nibble sequence, D5 AA 96, as well as some sector-specific metadata. (Email interview with author, February 16, 2021.)

27. Apple's first DOS, DOS 3.0, was released with the drive itself in June 1978. Its most long-standing version, DOS 3.3, came out in August 1980, quickly becoming an industry standard among Apple users. DOS 3.3 was a combination hardware/software upgrade; it came with a new System Master but also an improved ROM chip for the Disk II's controller card, which enhanced the operation of the Disk II drive without requiring expensive modifications to the drive itself. The new ROM chip, designed by Wozniak, used a more efficient encoding scheme that allowed Apple to increase the number of sectors on a disk from 13 to 16. Differences in the hardware and software design of floppy drives are why disks formatted for one hardware system, like the Apple II, cannot be read on another, such as the Commodore PET or the TRS-80. For detailed discussion of the internal structure of the Apple II DOS, see Weyhrich, *Sophistication*, 65–75; Worth and Lechner, *Beneath Apple DOS*.
28. For these exact instructions, see Apple Computer, *DOS Manual*, 38–40.
29. Numerous methods existed for copy protecting discs; see Chamberlain, "Apple II Copy Protection," for examples. The spiral technique, known as SpiraDisc, was created by a Sierra On-Line programmer named Mark Duchaineau. Levy, *Hackers*, 376–77. Elements of this explanation were also drawn from email conversations with 4am (see note 26).
30. Ekblaw, letter to the editor, 36.
31. Yuen, "Pirate, Thief," 14.
32. Tommervik, "Staggering Value," 17.
33. Morgan, "Editorial," 10.
34. Yuen, "Pirate, Thief," 15.
35. As stated, piracy was widely known to happen within user groups and other collectives, with varying degrees of organization. The drama surrounding Gates's "Open Letter to Hobbyists" (discussed in chap. 1) is just one indication of the prevalence of piracy, but discussions of the topic are routine in documents from the period. For examples, see Yuen, "Pirate, Thief"; Wollman, "Software Piracy." To a lesser extent, there was also concern that competitors might steal software code and market a product as their own. See Becker, "Legal Protection." Furthermore, the SPA materials in the Brøderbund Software, Inc., collection at the Strong Museum of Play illuminate a number of small- to medium-scale piracy operations (also discussed later in this chapter).
36. Harman, letter to the editor, 19.
37. A less dominant but still relevant reason for opposing copy protection was that it made it more challenging for users to access source code directly in order to fix bugs or customize the software to their needs. For documentation of some users who wanted to be able to edit source code, see Ekblaw, letter to the editor, 37.
38. Apple Computer, *DOS Manual*, 37.
39. Meehan, letter to the editor, 27; Anderson, letter to the editor, 35.
40. Fields, letter to the editor, 20.
41. A particularly illustrative example of this can be found in an April 1982 letter to the

editor in *Softalk* from a computer user who used *Locksmith* to break the copy protection on Sir-Tech's game *Wizardry*, after experiences "giv[ing] a command that requires *Wizardry* to access the disk and then never hear[ing] from the program again as the disk goes whisping around indefinitely." Behrens, letter to the editor, 8–9.

42. Milewski, "VCOPY Clones VisiCalc," 36.
43. Sirotek, letter to the editor, 22. According to Sirotek, 88 percent of the problems with damaged discs were due to information overwriting, which he defined as "data read back to a wrong area of a disk," an error typically caused by maintenance issues with the drive itself.
44. Brøderbund Crown, October 1984, Brøderbund Software, Inc. Collection, 37.
45. Con Díaz, *Software Rights*, 138. For more information on the deliberations that eventually led to the passage of the Computer Software Copyright Act in 1980, see Con Díaz, chap. 6.
46. Con Díaz, 137–38.
47. While *Copy II Plus* and *Back-It-Up* were generic bit copiers, *V-Copy* was solely designed to duplicate *VisiCalc* disks. See Milewski, "VCOPY Clones VisiCalc," 1, 36.
48. Alpert, "Censorship," 8.
49. Bayer, "No Truce in Sight," 17.
50. Omega Software Systems, "Locksmith," 80.
51. Omega Software Systems.
52. *Original Locksmith Users Manual*, 6.
53. Alpert, "Censorship," 8; *Original Locksmith Users Manual*, 6. The general air of mystery surrounding *Locksmith*'s provenance has been responsible for varied claims of the programmer's true identity, including that it was Alpert himself, or even Steve Wozniak. Alpert was named as the programmer in a 1985 *PC Magazine* article, while the Wozniak theory was circulated by the tech journalist Robert X. Cringely, who claimed he heard it from an Apple employee, Andy Hertzfeld. Lewenstein, "Ethics," 186; Cringely, "Verizon's iPhone Story." However, scattered documentation suggests a man named Mark Pump may have been *Locksmith*'s anonymous author. In both cases, responses to the articles in question corrected the name of the programmer to Mark Pump: first, a letter Alpert wrote to *PC Magazine* noting the inaccuracy in attributing *Locksmith* to him; and second, a comment on Cringely's website left by someone named "Mike" who claimed to have worked with Pump at Omega. Alpert, letter to the editor, 63. Neither source can be confirmed. Neither Pump nor Alpert seems to have retained any role in the personal computing software industry beyond the mid- to late 1980s; Omega MicroWare, which does not appear to have existed before 1981, similarly vanishes from the record by the mid-1980s. However, Pump appears in a February 1985 episode of *Computer Chronicles* as the president of a company called Alpha Logic, which was distributing later versions of *Locksmith*. My own efforts to locate either Dave Alpert or Mark Pump did not turn up leads.
54. Lock, "Editor's Notes," 4, 9.
55. Lock, 9.
56. "No More!," 18.
57. Response to these editorials was fairly mixed and in some cases inspired further comment from editorial staff. *InfoWorld* received two letters, both of which supported the right of the consumer to use bit copier software. See Naritomi, letter to the editor, 13, 47; Leavitt, letter to the editor, 12. In its May 1981 issue, *COMPUTE!* published an uncommonly extensive reply that came from the affinity group Com-

puter Using Educators (CUE), which submitted an entire position paper to address the need for schools to be able to easily and affordably duplicate their software to protect against student mishandling and excessive cost. In July 1981, *COMPUTE!* would publish a follow-up editorial on software copyright, aimed at correcting reader inaccuracies on the subject.

58. *Call-A.P.P.L.E.* also made an editorial pledge against bit copy ads. Bayer, "No Truce," 17. *Softalk*, while not making an explicit statement on the subject, reportedly returned Omega's advertising money, "with interest." Alpert, "Censorship," 9.

59. *Hardcore* gained most of its initial 200 subscribers by advertising itself somewhat under the radar in *Softalk* and *Call-A.P.P.L.E.* The advertisement emphasized *Hardcore*'s focus on copy protection and backup disks without ever explicitly stating it was in favor of breaking copy protection. See advertisement in *Softalk*, May 1981, 70. Also see "Thanks," 11.

60. Haight, "What I Need," 2.

61. Haight.

62. Haight.

63. Haight, "Censorship," 4. While Chuck and Bev share a last name, their exact relationship is not clear. While it is tempting to assume Bev is short for Beverly (as Con Díaz does in the first printing of *Software Rights*), Chuck uses the masculine pronoun *he* to refer to Bev in an editorial from the second issue. The media scholar Gavin Mueller claims they are brothers, though that claim is not evidenced (*Media Piracy*, 58). However, we can be fairly certain Chuck was not married to Bev. According to county records from Tacoma, Washington (where *Hardcore* was published), Charles Haight married Karen Fitzpatrick in 1981; Karen Fitzpatrick is also the name of *Hardcore*'s subscription manager. My thanks to Con Díaz for investigating county records when I followed up with him about the relationship between Charles and Bev.

64. Fitzpatrick, "Bit Copy Programs," 10.

65. Alpert, "Censorship," 8.

66. Val Golding's role in the early microcomputing enthusiast magazine and user group world was multilayered. *Call-A.P.P.L.E.* was a publication of the Apple Puget Sound program library exchange—what was ostensibly a fairly polished user group magazine that had wide circulation beyond that specific geographic region. Golding was also the former editor of *Apple Orchard*, a publication of the International Apple Core (IAC), an organization founded to help coordinate and share resources between various Apple user groups. For a brief discussion of *Apple Orchard* and the history of the IAC, see Weyhrich, *Sophistication*, 175–76.

67. Alpert, "Censorship," 8.

68. Alpert.

69. Alpert, 9.

70. Alpert.

71. Bayer, "No Truce," 17.

72. Golding, "Rebuttal," 12. Alpert's accusations were also not constrained to the pages of *Hardcore*. He wrote a letter to *Softalk* accusing the magazine of hypocrisy for running ads for On-Line Systems' erotic adventure game *Softporn* but refusing to run *Locksmith* ads. Alpert, letter to the editor, *Softalk*, 10.

73. Alpert, letter to the editor, *Softalk*, 9.

74. *Locksmith* first appears in the February 1982 Hobby Top 10 list, stays there through April, and falls off in June (it appears *Softalk* forgot to publish the May list).

75. "Most Popular Program," 166.
76. According to *Softalk*'s own rules, however, *Locksmith* wasn't even supposed to be eligible for inclusion, since the goal of the awards was to highlight new programs, and *Locksmith* had been released a year earlier, in 1981. The confusion came from the fact that *Locksmith*, like several other programs that were included (such as *Screenwriter II*, *PFS: File*, *Apple Writer II*, and *Global Program Line Editor*) had undergone significant revision the previous year and was therefore placed on the list, even though *Softalk* had disqualified other software for the same reason in prior years. Reluctant to disqualify thousands of votes they received, *Softalk* "hit upon a solution certain to be unsatisfactory to everyone," ranking all programs according to the number of votes they received but putting an asterisk rather than a numbered rank next to those programs that were ineligible. "It's *Choplifter*," 77.
77. As examples, advertisements for these products can be found in *MICRO 6502*, July 1981, 86; *Softalk*, November 1981, 59; *Softalk*, September 1982, 91. On MobyGames, the full title of the game is listed as *A City Dies Whenever Night Falls*; see https://www.mobygames.com/game/a-city-dies-whenever-night-falls. For a review of this game, see Harrington, "Marketalk Reviews," 155.
78. This advertisement ran in the January 1982 issue of *Creative Computing*, the November 1981 issue of *BYTE*, and two issues of *Kilobaud Microcomputing*, the earliest of which was October 1981.
79. Morgan, "Editorial," 6.
80. Pelczarski, letter to the editor, 26.
81. Pelczarski, 26–27.
82. Neiburger, letter to the editor, 26. Neiburger, an advocate for bit copiers, also appears in *InfoWorld*'s issue dated May 25, 1981, where we learn that he is a dentist from Waukegan, Illinois, who also published the *Dental Computer Newsletter*. Responding to Robert Tripp's *MICRO* editorial in the pages of his own newsletter, Neiburger suggested perhaps readers should boycott *MICRO*'s advertisers to "counter the pressure he felt sure advertisers were placing on the magazine." Bayer, "No Truce," 18–19. Neiburger appears to have been a significant player in introducing personal computers into dentistry. He was the "Dental SIG" chairman for the International Apple Corps and a consultant for Apple Computer. See "Contributors," *Chairside Magazine* 6, no. 1 (January 31, 2011), https://glidewelldental.com/education/chairside-dental-magazine/volume-6-issue-1/contributors/; http://www.drneiburger.com/dr.-ns-books---articles.html; http://www.drneiburger.com/about-us.html.
83. Software Publishers Association, "Open Letter"; Doug Carlston to the SPA membership, March 13, 1985, Brøderbund Software, Inc. Collection.
84. "SPA Copyright Protection Fund," SPA internal documentation, July 25, 1985, Brøderbund Software, Inc. Collection.
85. "SPA Copyright Protection Fund."
86. "Future Visions Computer Store," July 25, 1985, Brøderbund Software, Inc. Collection.
87. John Kim to SPA, January 5, 1985, Brøderbund Software, Inc. Collection.

Chapter 6

1. Ad found on p. 14 of the issue.
2. What appears to be an identical framed photograph can also be seen in a handful of

Getty images taken at the Faire—a detail that seems plausible given Regis McKenna designed both Apple's inaugural booth and this advertisement.

3. This number comes from Sigel and Giglio's 1984 market report, *Guide to Software Publishing*, table 1.6, p. 18. Sigel and Giglio's work divides the microcomputer market into three categories: personal (i.e., workplace, with an installed base of 2.7 million), home (installed base of 6 million), and education (installed base of 400,000). However, in the case of personal and home computers, these two categories are defined by the brand of microcomputer, not by where the computer was used. Due to its high price point, Apple is classified as a personal computer, not a home computer. Other manufacturers, like Tandy, appear in both categories. Therefore, it is impossible to know for certain what percentage of microcomputers in the personal computer category were actually being used at home. It seems plausible that an actual estimate of computers in the home, based on Sigel and Giglio's statistics, could be higher than 6 million.
4. Gutman, "Praising Weirdware," 7.
5. Gutman.
6. Willis and Miller, *Computers for Everybody*.
7. Goehner, "Pixellite Software," 34.
8. Exceptions prove the rule: there are documented cases of families with teletype terminals in their homes, beginning in the 1960s, see Rankin, *People's History*, 78–79. On a national scale, however, such use would have been exceedingly rare.
9. Spicer, "ECHO IV"; *Neiman-Marcus Christmas Book*, 84. Also see Marcus, *His and Hers*, 123–30; Atkinson, "Curious Case." For coverage of ECHO IV, see Infield, "A Computer in the Basement?"; Spicer, "If You Can't Stand the Coding"; "Electronic Computer for Home Operation."
10. Salsberg, editorial, 4.
11. One issue included a speculative essay on the future of home computing for financial planning, dietary management, systems control, and personal entertainment. Gardner, "The Shadow."
12. Haller; "Golf Handicapping"; Fox and Fox, "Biorhythm for Computers."
13. Lau, "Total Kitchen Information System," 45.
14. Ciarcia, "Computerize a Home," 28.
15. "Softalk Presents the Bestsellers," *Softalk*, October 1980, 27. As the software market developed, word processors and typing instruction software tended to be treated as their own software subtype.
16. This was also the first issue that *Softalk* provided a dedicated "Business" top ten list.
17. The one caveat to this would be that *Softalk*'s best sellers only reflected retail sales; software sold primarily by mail order was not accounted for in these metrics.
18. "Top Thirty Bestsellers," *Softalk*, September 1981, 116.
19. "Top Thirty Bestsellers," *Softalk*, October 1981, 145–47.
20. "Top Thirty Bestsellers," 147.
21. Nooney, Driscoll, and Allen, "From Programming to Products," 119.
22. Nooney, Driscoll, and Allen.
23. Nooney, Driscoll, and Allen, 122–23.
24. Williams, "Computers," 243.
25. See Nelson, *Computer Lib/Dream Machines*; Levy, *Hackers*, 39–51.
26. Tommervik, "Straightalk," 3. For more information on *Softalk*'s unique editorial approach, see Nooney, Driscoll, and Allen, "From Programming to Products."
27. *Popular Computing* was a rebranding of McGraw-Hill's *onComputing*, a bimonthly

magazine for computing beginners that it had launched in 1979. Morgan, "Over into the Future," 4.

28. Cohl, "Join Us," 8.

29. Reed, "Domesticating," 170. For more information on the prevalence of computer-phobia in the personal computing era, see 169–75.

30. Reed, 174.

31. Willis and Miller, *Computers for Everybody*, 2–3.

32. Kortum, "Confessions," 36.

33. In science, technology, and society (STS) literatures, such processes might be variously referred to as "stabilization" in the sense of Wiebe Bijker or a "configuration of users" according to Stephen Woolgar. See Bijker, Hughes, and Pinch, *Social Construction*; Woolgar, "Configuring the User."

34. Zonderman, "From Diapers to Disk Drives"; Ball, "The Computer."

35. See Olson, "Family Computer Album"; Levine et al., "Family Computer Diary."

36. Levine et al., "Family Computer Diary," 117.

37. Willis and Miller, *Computers for Everybody*, 61–71.

38. Willis and Miller, 61.

39. Eisenhart, "Broderbund Software," 40.

40. David Balsam, Zoom interview with author, June 6, 2021. According to Balsam's recollection, the school was equipped with both an IBM 360 and a lab of Olivetti and Wang desktop systems, which he referred to as "microcomputers." While Balsam did not identify the specific model numbers, it is probable that these were the Wang 2200 and possibly the Olivetti P602, both of which were released in the early to mid-1970s. Both systems were oriented to business and research functions and never entered the consumer market. As dedicated, general-purpose, desktop-sized computing systems, they are part of a class of systems that complicate straightforward genealogies between minicomputers and microcomputers. As an anecdotal observation, I have never previously heard of a school system with a computer lab that used anything besides time-sharing or punch cards that early in the 1970s. If Balsam's recollections are correct, the computer lab at John Dewey was a unique site.

41. Martin Kahn, Zoom interview with author, June 10, 2021.

42. Kahn tried to break into the San Francisco gallery scene making computer-generated art, but found there was little interest in the medium. However, he did pull together a collection of his black-and-white works in a coloring book called *Computer Imagery*, published by Golden Triangle Distributors in 1981.

43. Goehner, "Pixellite Software," 32.

44. Kahn, interview.

45. Kahn, interview.

46. Balsam, interview. According to Brøderbund's internal newsletter, *The Brøderbund Crown*, 79 percent of employees were under 35 years old. *The Brøderbund Crown* 1, no. 2, June 1983, Brøderbund Software, Inc. Collection, 9.

47. For a fairly comprehensive first-person account of the founding of Brøderbund, see Carlston, *Software People*, 57–71.

48. Carlston, *Software People*, 6.

49. Explanations for the origin of the name are mixed, often citing it as Swedish due to Gary Carlston's Scandinavian Studies degree and the time he spent in Sweden coaching women's basketball. Doug also spent time in Botswana, where he would have likely encountered the Dutch colonial language Afrikaans, which shares West Germanic roots with various Scandinavian languages. The "ø" does double duty as both a Danish vowel and a reference to the slashed zero glyph used in ASCII to distinguish the numeral from its alphabetic counterpart. Doug Carlston, interview by Tim Bergin, November 19, 2004.

50. Campbell-Kelly, *Airline Reservations*, 225.
51. Brøderbund had honed this industry positioning very early in its corporate growth as a consequence of becoming the US distributor for the Japanese software development company Star Craft in 1981.
52. Balsam, interview.
53. Goehner, "Pixellite Software," 32.
54. Goehner.
55. Both Balsam and Kahn recall that they showed the prototype to Gary Carlston, who, according to company records, was the vice president of product development at the time. *The Brøderbund Crown* 1, no. 1, May 1983, Brøderbund Software, Inc. Collection, 6.
56. Goehner, "Pixellite Software," 34. The lack of scalability would have been exacerbated by the fact that both sender and recipient would have needed to have *the same* microcomputer, since floppy disk formatting was platform-dependent.
57. Rosch, "Printer Comparisons"; Ahl, "Buying a Printer"; Mazur, "Hardtalk."
58. Rosch, "Printer Comparisons," 35.
59. Ahl, "Buying a Printer," 12.
60. For more information about Apple printers and Apple II–compatible printers, see Weyhrich, "Bits to Ink," in *Sophistication*, 231–37.
61. Balsam, interview.
62. Cronk and Whittaker, *PrintShop Reference Manual*, 13.
63. Balsam, interview; Kahn, interview; Cosak, interview.
64. According to Balsam, they also pitched to Electronic Arts but mostly only to gain leverage with Brøderbund, which they knew they wanted to work with. Doug Carlston recollects that the revised prototype was brought back to Brøderbund in September 1983. Eisenhart, "Broderbund Software," 42.
65. For more information on the features and development history of the Apple IIe, see Weyhrich, *Sophistication*, 197–201.
66. Willis and Miller, *Computers for Everybody*, 191.
67. Goehner, "Pixellite Software," 34. Balsam originally recalled royalties of 30 percent, but Kahn corrected this to 20 percent. Balsam, interview; Kahn, interview.
68. Martin Kahn stated, in an email, that the program's name was his idea. I did not have the opportunity to cross-reference this with Brøderbund corporate materials or Doug and Gary Carlston.
69. Goehner, "Pixellite Software," 34.
70. Not all modes had access to the same design elements. For example, banners and letterhead could not be printed with a border, letterhead could not have address information in outline form, etc. Furthermore, *The Print Shop* also provided a mode called "Screen Magic," which is essentially the original prototype of Perfect Occasion.
71. While *The Print Shop* is sometimes referred to as a form of early desktop publishing software, it carried none of the drag-and-drop or WYSIWYG functionality that defined programs like *PageMaker* or *QuarkXPress* when they emerged in the late 1980s.
72. Goehner, "Pixellite Software," 35; emphasis in original.
73. Tommervik, "The Print Shop," 114; emphasis in original.
74. Tommervik, 115.

75. McCollough, "The Print Shop," 74.
76. Goehner, "Pixellite Software," 32.
77. Zuckerman, "New on the Charts," 31.
78. Carlston, Foreword to *The Official Print Shop Handbook*, xiii. Brøderbund staff even used the program to design the company's internal newsletter, *The Brøderbund Crown*, which can be found in the company archives at the Strong Museum of Play.
79. *The Print Shop* did appear in *Softalk*'s last top thirty best seller list in August 1984; it ranked twentieth overall and third in home software. For coverage of *Billboard*'s introduction of the computer software chart, see "New Charts for Computers, Vidisks," 3, 76. The first computer software *Billboard* chart also appears in this issue, offering an entertainment top twenty, an education top ten, and a home management top ten. The most definitive source for microcomputer bestseller information would be the distributor Softsel's weekly "Hotlist," but no archive or collection of this information is known to exist at this time.
80. While Kahn programmed the drivers for Apple II–compatible printers, the Brøderbund programmer Corey Kosak did the ports for Commodore and Atari. Corey Kosak, Zoom interview with author, May 30, 2021.
81. Doug Carlston, phone interview with author, April 7, 2021.
82. "Inside the Industry," 8.
83. "Inside the Industry."
84. Benton, Author's Note, in *The Official Print Shop Handbook*, xv.
85. Benton.
86. For information on the Unison v. Brøderbund lawsuit, see "Unison Loses Software Suit," *New York Times*, October 16, 1986, D22. To identify which program made this banner, I cross-referenced the images and fonts against the manuals for both programs. The flower images depicted on the printout are the "rose" and "bouquet" art available in *PrintMaster*. https://archive.org/details/Print_Master_Users_Guide/.
87. Claim based on MobyGames' history of Brøderbund logos, though a general YouTube search of Brøderbund logos suggests there was some use of an unslashed *o* within some studios beginning in the mid-1990s. See https://www.mobygames.com/company/brderbund-software-inc/logos.
88. Balsam and Kahn had been joined at this point by Ken Grant, who became a third partner at Pixellite. In Balsam's words, he was "instrumental in managing the growing software team, as well as negotiating contracts with other publishers." Balsam, email with author, March 28, 2022.
89. Maxis, "Maxis Print Artist Advertisement (1995)."

Chapter 7

1. David Ahl provides a first-person account of the magazine's founding in "Birth of a Magazine."
2. The Apple Macintosh was not Apple's first computer with a graphical user interface. That distinction goes to the Apple Lisa, released in 1983. However, the Lisa was not aimed at consumers or even small businesses. Priced at roughly $10,000, the computing unit was understood as a "workstation" computer—something you would find only in an office.
3. Staples, "Educational Computing," 62.
4. These numbers derive from Reed, "Schools Enter the Computer Age"; "Micro Use Is up"; Chion-Kenney, "Schools Bought." Quote comes from Chion-Kenney. Statis-

tics were based on a national survey conducted by the private market research firm Market Data Retrieval.

5. Watt, "Computers in Education," 83.
6. Chion-Kenney, "Schools Bought."
7. Kay Savetz, Interview with Bill Bowman, https://archive.org/details/bill-bowman-320/Bill+Bowman+320.mp3.
8. Tom Snyder, Zoom interview with author, July 25, 2021. The January 1978 date comes from Ahl, "Profile."
9. Snyder, interview.
10. Snyder.
11. Snyder.
12. Birkel, "Leadings."
13. Snyder originally called the device the "Consensor," a portmanteau of "consensus" and "sensors." He explains that changing the name to Personk was something of a private joke between himself and Waddington: "At one point I had as many as 'K,' A-through-K, people be able to plug into it at once. . . . My girlfriend [Anne Waddington] came in and said 'What's Personk?' . . . because I had written 'Person K' on a little controller, and I said, 'That's the name of my product.'" Snyder, interview.
14. Snyder, interview.
15. "Memorial."
16. McGraw-Hill, annual reports, 1978–82.
17. McGraw-Hill, *1979 Annual Report*, 10; McGraw-Hill, *1980 Annual Report*, 12.
18. For more reading on the history of educational computing, particularly in the United States, see Rankin, *People's History*; Cain, *Schools and Screens*; Boenig-Liptsin, "Making Citizens." Texts relevant to a broader history of educational computing efforts include Ames, *Charisma Machine*; Cuban, *Oversold and Underused*. For a more general study of how technology was used to mechanize and automate teaching, especially at midcentury, see Watters, *Teaching Machines*.
19. Cain, *Schools and Screens*, 143.
20. Minnesota's MECC network—which, with 84 percent of the state's public school students participating by 1975, was the largest educational time-sharing network in the country—exemplifies the extent to which such educational implementations required sustained political will and advantageous civic conditions. The capacity of Minnesota school districts to form their own independently managed time-sharing network was a direct consequence of the state's joint powers authority, which permitted individual districts to band together as a single entity to bid for time-sharing services.
21. Reed, "Schools Enter the Computer Age," 1.
22. Cain, *Schools and Screens*, 144.
23. As Cain notes, the individualized qualities of microcomputers appealed to both "countercultural emphasis on individuality and creativity" as well as conservative fixations on "personal autonomy and achievement over collective progress." Cain, 146.
24. Cain, 145.
25. Snyder, interview.
26. Each of the five programs was sold individually; the series included *Geography Search*, *Energy Search*, *Community Search*, *Archaeology Search*, and *Geology Search*. To read a review of the *Search Series*, see Boston, "Search Series." For descriptions

of the software from teaching resources, see Hunter, *My Students*, 198–200; Braun, *Microcomputers*, 108–9.

27. McGraw-Hill, *1981 Annual Report*, 7; California State Board of Education, *Program Descriptions*, 19.
28. It's unclear when the hardware manufacturers of the 1977 Trinity began offering educational discounts, but by 1981 it was reported that Apple offered a 5 percent discount, RadioShack offered a 10 percent discount, and Commodore had a "3 for 2" deal for educational buyers. See Grady and Gawronski, *Computers*, 49.
29. Rankin, *People's History*, 238–40.
30. Rankin, 240.
31. In June 1981, the *Wall Street Journal* reported that TA Associates had the highest rate of paid-in capital of any independent firm in the country at $105 million ($5 million more than its nearest competitor). "Congress and Estate Taxes," 23.
32. "Spinnaker Software Corporation," 1.
33. Zientara, *Women*, 131.
34. For brief biographical treatments of Morby, see Zientara, *Women*, 127–34; Zientara, "Five Powerful Women," 58.
35. "Digital Research," 12.
36. Nicholas, *VC*, 177; Pierson, "Steiger's Capital Gains," 20; "Blumenthal," 3. Lower capital gains taxes not only allowed individual investors to hold onto more of their profits, but were also believed to stimulate entrepreneurial activity by making it less risky for potential entrepreneurs to leave waged work. For more on the impact of capital gains on venture capital, see Nicholas, *VC*, 177–81.
37. See Nicholas, *VC*, 173–77.
38. Zientara, *Women*, 127.
39. Nicholas, *VC*, 178.
40. "Spinnaker Software Corporation," 3.
41. For example, Brøderbund's marketing was headed by Doug and Gary Carlston's sister Cathy, who did have some retail training as a former buyer for Lord & Taylor. A more typical example resides with Sierra On-Line, where marketing, such as it existed, was run by Ken Williams's younger brother, Johnny, who had no prior experience.
42. "Spinnaker Software Corporation," 3.
43. "Spinnaker Software Corporation," 4.
44. For a robust analysis of Seymour Papert's development of LOGO, and its ideological underpinnings, see Ames, *Charisma Machine*. For the history of PLATO, see Rankin, *People's History*, chaps. 6 and 7. The prehistory of computers in education, focused on mechanized educational machines, can be found in Watters, *Teaching Machines*. Additionally, a great deal of activity in research happened under the awning of what was called "computer-aided learning," which is heavily associated with the research of Stanford professor Patrick Suppes, among others.
45. The founders of The Learning Company and Edu-Ware, both prominent educational software developers and publishers of the 1980s, both shared pedagogic backgrounds. Edu-Ware's founder, Sherwin Steffin, was an educational technologist. Teri Perl and Ann McCormick of The Learning Company both had PhDs in education. See Tommervik, "Exec Edu-Ware"; CRAW, "Teri Perl"; "Ann McCormick."
46. For a good overview of the breadth of educational software development and publishing in the very early 1980s, see Lubar, "Educational Software."
47. We know this due to the relative absence of such advertising in popular computer

enthusiast magazines of the period. However, products from these companies do routinely appear in educational software directories of the mid-1980s.

48. "Spinnaker Software Corporation," 6.
49. "Spinnaker Software Corporation," 4.
50. Zientara, *Women*, 131–32.
51. "Spinnaker Software Corporation," 6–7. According to Exhibit 1 from this document, a reproduction of Spinnaker's "Critical Path" chart from June 22, 1982, work on Spinnaker's logo began May 14, and it was approved on May 28.
52. Snyder, interview; Bowman, interview.
53. Snyder, interview. The president of CLC was named Rick Abrams.
54. Exactly what this piece of software did is unclear. In Bill Bowman's interview, he states that Morby's son had problems with writing. In my own interview with Snyder, he described the program thus: "I wrote a side program that put the text up, and as you typed—this was on my TRS-80—it would tell you when you made a mistake. And you could set it, as a student, for how long after you've made a mistake, that you'd want it to tell you. So it was just a little slow word processor I built that would take his homework, or whatever, and a parent could type it in, or I could type it in this case, and then I could, and let him play with it."
55. Snyder, interview.
56. "Spinnaker Software Corporation," 4, 19.
57. Snyder, interview.
58. Such techniques were already somewhat common within certain forms of game development, especially textual and graphical adventures. This was a goal Snyder had aspired to but never quite achieved with *The Search Series*.
59. The stories for *Snooper Troops #1* and *#2* were written by Karen Eagan, and Deborah Kovacs and Patricia Relf, respectively. Kovacs and Relf were both former employees at the Children's Television Network (makers of *Sesame Street*) and took up Snyder's project as a freelance gig.
60. Cleverly, the game's win state can be achieved only by *ruling out* individual suspects using specific clue data (thus preventing random guessing).
61. Snyder, interview.
62. While Spinnaker's launch lineup featured four games, the company had initially reported that it would be offering ten, according to an industry blurb in *Softalk*'s July 1982 "Tradetalk" section. DesignWare was founded by Jim Schuyler in 1980. For more on the company's history and Schuyler's background, see Schuyler, "DesignWare's Founding"; Schuyler, "Jim Schuyler."
63. The cases were manufactured by a company in Minneapolis. Bowman, interview.
64. Most software was sold on cassette or floppy disk, and packaged loose in a box along with a variety of print materials that came in numerous shapes and sizes, including product catalogs, game manuals, additional instructions, warranty cards, and promotions for other companies. In 1983, Spinnaker's cases would be slightly redesigned to accommodate cartridges, with the product casing becoming taller and the dimensions on the instructional materials becoming more horizontal than vertical. These changes do not appear to have impacted disk-based product packaging.
65. Several documented accounts illustrate how unprepared Bowman and Seuss were to run a microcomputer software company. For example, in Bowman's interview with Savetz, he discusses the surprise they had realizing they would need a warehouse for storing their manufactured goods and workers to assemble them. Also see "Spinnaker Software Corporation," 5.

66. These advertisements can be found in *Creative Computing*, November 1982, 16–17; *COMPUTE!*, October 1982, 4–5; *PC Magazine*, September 1982, 14–15; *Softalk*, September 1982, 12–13.
67. Tommervik, "Marketalk Reviews."
68. Ahl, "Learning," 134.
69. Fiske, "Schools."
70. Fiske, 38.
71. For examples, see Toffler, *Future Shock* and *The Third Wave*. Papert was the inventor of the child-friendly programming language LOGO, which was widely circulated in both educational and consumer spaces during this time. Luehrmann is associated with the popularization of computer literacy; see his set of essays in Taylor, *Computer in the School*, 127–58.
72. A good example demonstrating the general inaccessibility of first-generation microcomputing for children's educational needs can be found in *BYTE*'s November 1976 cover story, "It's More Fun than Crayons." Written by a microcomputer hobbyist with five- and seven-year-old sons, the feature documents the children's experience using their father's 1 KB homebrewed microcomputer to create simple graphical images on the system's television screen output. The father acknowledged this took the seven-year-old "about two hours of instruction," while the younger boy required continual assistance. In such a scenario, the involvement of the hobbyist father is often very much the point—emphasizing the literal passing down of personal interests along gendered and generational lines—but it also indicates how little the computer itself did for children without the support of a committed hobbyist already in the household. Rosner, "It's More Fun," 9.
73. For examples of early advertisements for educational software, see *BYTE*, March 1979, 127; *Creative Computing*, November 1979, 11, 40.
74. Some of this might be explained on the assumption that educators formed a not inconsiderable portion of the reader base of computer enthusiast magazines, but it's also the case that the educational publishing industry fairly quickly started releasing guidebooks and catalogs of software for schools.
75. Lubar, "Educational Software," 64.
76. While this was the first time education software had reared its head as an issue of categorization, *Softalk* had launched a new section called "The Schoolhouse Apple" just a few months earlier, in May 1982. This section was aimed at "mak[ing] it easier for you to find your way around in the seeming labyrinth, to grasp the essentials amid the overload of this aspect of the information explosion." While the new section acknowledged that the educational computing field encompassed software developed for "schools, homes, and businesses," the section largely catered to information focused on the applications of computers in schools, as well as company listings. Varven, "Schoolhouse Apple," 36.
77. "*Softalk* Presents the Bestsellers," December 1982, 336. Classifying *Snooper Troops* as a "fantasy role-playing game" relies on a generous interpretation of both genres. The game is a "role-playing" game only in the sense that players adopt the role of a detective (there are no stats or character attributes managed by the player, which is a common characteristic of the RPG). The game is a "fantasy" only in the sense that it is set in a fictional, though very American, East Coast town.
78. "*Softalk* Presents the Bestsellers," *Softalk*, January 1983, 249.
79. While children could certainly use these programs, especially *MasterType*, which used the artifice of a space shooter to run typing drills, the programs came with no

suggested age range. Given the popularity of typing instruction software, it is plausible a good portion of the consumer base was adult men who had never learned touch typing (a feminized labor skill) in school. In the June 1983 issue of *Softalk*, the editors comment on the trend they were seeing of new buyers simultaneously purchasing word processors and typing instructional software. "*Softalk* Presents the Bestsellers," *Softalk*, June 1983, 281.

80. The only other software program on the list that reasonably also addressed adults was *Apple LOGO*, though the programming language was largely considered to be developed for children.
81. "*Softalk* Presents the Bestsellers," *Softalk*, November 1983, 335–36.
82. "Summer-CES Report," 35.
83. "Spinnaker Software Corporation," 6.
84. For evidence of these advertisements, see: *Good Housekeeping*, October 1983, 127; *Better Homes & Gardens*, October 1983, 54; *Newsweek*, October 31, 1983, 50.
85. *Better Homes & Gardens*, October 1983, 54.
86. "Spinnaker Software Corporation," 6.
87. *Billboard* initially tracked sales across Apple II, Atari, Commodore, IBM, Texas Instruments, TRS, CP/M, and "other." In 1984, Texas Instruments was removed after the shuttering of their microcomputing division, and the category was taken over by the Macintosh.
88. Spinnaker's most popular products at this time were *Facemaker* and *Kindercomp*.
89. The $50 million valuation was based on raising an additional $5 million during what was supposed to be a mezzanine round—a form of medium-risk capital investment that traditionally precedes an IPO. For a deeper explanation of the function of mezzanine funding, see Nicholas, *VC*, 241–42.
90. "Spinnaker Software Corporation," 7.
91. "Spinnaker Software Corporation."
92. Spinnaker struck a deal where they used Fisher-Price's name and produced software for them. "Spinnaker Software Corporation," 10–11.
93. National Commission on Excellence in Education, *Nation at Risk*.
94. "*Softalk* Presents the Bestsellers," *Softalk*, May 1984, 208.
95. Rosenberg, "Bowman," 93.
96. Rosenberg.
97. Rosenberg, "*A* Is for Acquisition," 41.
98. Rosenberg.
99. "Scholastic Acquires Tom Snyder Productions."
100. Cuban, *Oversold and Underused*, 196.

Epilogue

1. Sammis, "Wireless Girdle," 255.
2. Sammis.
3. This location in New Jersey is in Lenapehoking, the unceded territory of the Lenape peoples, which extends from New York to New Jersey, Delaware, and sections of eastern Pennsylvania.
4. InfoAge, "Visitor Self-Tour Map."
5. InfoAge.
6. For more information on the festival and its organizing institution, see Vintage Computer Federation, "Vintage Computer Festival East."

7. To get a glimpse of the scale of Scott's uploading work, see Scott, "User Account." To explore his personal website, which includes his prominent collection of BBS text files, see Scott, textfiles.com.
8. Nooney, "Pedestal"; emphasis in original.

Bibliography

Abbate, Janet. *Recoding Gender: Women's Changing Participation in Computing*. Cambridge, MA: MIT Press, 2012.

Ahl, David H. "Birth of a Magazine." *Creative Computing*, March–April 1975, 6–7.

———. "Buying a Printer." *Creative Computing*, March 1983. https://archive.org/details/CreativeComputingbetterScan198303/.

———. "Computer Power to the People! The Myth, the Reality, and the Challenge." *Creative Computing*, May–June 1977.

———. "Dave Tells Ahl." Interview by John Anderson. *Creative Computing*, November 1984.

———. "The First West Coast Computer Faire." *Creative Computing*, July–August 1977.

———. "Learning Can Be Fun." *Creative Computing*, April 1983.

———. "Profile of a Super Trooper." In *Creative Computing Software Buyers Guide 1983*, 60. Morris Plains, NJ: Ahl Computing, 1983.

Albert, Dave, Douglas G. Carlston, Margot Comstock, Jerry W. Jewell, Ken Williams, and Roberta Williams. Interview by Jon B. Eklund,

July 31, 1987. Smithsonian Institution Archives, Record Unit 9533 Minicomputers and Microcomputers Interview.

Alcorn, Allan. Interview by Henry Lowood, April 26, 2008, and May 23, 2008. "Oral History of Allan (Al) Alcorn." Computer History Museum. https://archive.computerhistory.org/resources/access/text/2012/09/102658257-05-01-acc.pdf.

Allen, Kera. "'The Official Response Is Never Enough': Bringing VisiCalc to Tunisia." *IEEE Annals of the History of Computing* 41.1 (2019): 34–46.

Allen, Paul. *Idea Man: A Memoir by the Cofounder of Microsoft*. New York: Penguin, 2012.

Allison, Dennis, Happy Lady, and Friends. "DESIGN NOTES FOR TINY BASIC." *People's Computer Company* 4, no. 2 (September 1975): 15–18. https://archive.computerhistory.org/resources/access/text/2017/09/102661095/102661095-05-v4-n2-acc.pdf.

Almaguer, Tomas. *Racial Fault Lines: The Historical Origins of White Supremacy in California*. Berkeley: University of California Press, 2008.

Alpert, David M. "Censorship in Computer Magazines, part 2: An Interview with: Dave Alpert of Omega Software about the 'Locksmith' Ad. Controversy." Interview by *Hardcore Computing* magazine. *Hardcore Computing* 1, no. 1 (1981): 8–9. https://archive.org/details/hardcore-computing-1/.

———. Letter to the editor. *PC Magazine*, April 1985.

———. Letter to the editor. *Softalk*, December 1981.

Altair Users Group Software Library. Atlanta, GA: Altair Software Distribution Company, [1977?]. Vintage Technology Digital Archive, http://vtda.org/docs/computing/AltairUserGroup/AltairUserGroup_SoftwareLibraryCatalog.pdf.

Amankwah-Amoah, Joseph. "Competing Technologies, Competing Forces: The Rise and Fall of the Floppy Disk, 1971–2010." *Technological Forecasting and Social Change* 107 (June 2016): 121–29.

Ames, Morgan G. *The Charisma Machine: The Life, Death, and Legacy of One Laptop per Child*. Cambridge, MA: MIT Press, 2019.

Anderson, Tim. Letter to the editor. *Softalk*, May 1983.

"Ann McCormick." Dust or Magic. Accessed December 31, 2021. https://dustormagic.com/speaker/ann-mccormick/.

Apple Computer. *The DOS Manual: Apple II Disk Operating System*. Cupertino, CA: Apple Computer Inc., 1980. https://archive.org/details/the-dos-manual/.

Arthur, W. Brian. *The Nature of Technology: What It Is and How It Evolves*. New York: Free Press, 2009.

Atkinson, Paul. "The Curious Case of the Kitchen Computer: Products and Non-Products in Design History." *Journal of Design History* 23, no. 2 (2010): 163–79.

Aycock, John. *Retrogame Archaeology: Exploring Old Computer Games*. Dordrecht: Springer, 2016.

Ball, Jeff. "The Computer: A New Tool for the Garden." *Family Computing*, March 1984, 38–41.

Barak, Sylvie. "John McCollum: HS Teacher Taught Electronics with Love." *EE Times*, December 11, 2012. https://www.eetimes.com/john-mccollum-hs-teacher-taught-electronics-with-love/.

Bardini, Thierry. *Bootstrapping: Douglas Engelbart, Coevolution, and the Origins of Personal Computing*. Stanford, CA: Stanford University Press, 2000.

Bates, William. "The Floppy Disk Comes of Age." *New York Times*, December 10, 1978, F11.

Bayer, Barry D. "No Truce in Sight in Copy-Protection War." *InfoWorld*, May 25, 1981.

Becker, Stephen A. "Legal Protection for Computer Hardware and Software." *BYTE*, May 1981.

Beeler, Jeffry. "Personal Computing Is Big Business: No Turning Back." *Computerworld*, December 27, 1982–January 3, 1983, 21–24. https://archive.org/details/sim_computerworld_december-27-1982-january-03-1983_16_52/.

Behrens, Adam. Letter to the editor. *Softalk*, April 1982.

Benton, Randi. Author's Note, in *The Official Print Shop Handbook: Ideas, Tips, and Designs for Home, School, and Professional Use*, by Randi Benton and Mary Schenck Balcer, xv–xvi. New York: Bantam Books, 1987. https://archive.org/details/officialprintshoooobent/.

Bijker, Wiebe E., Thomas Parke Hughes, and Trevor Pinch, eds. *The Social Construction of Technological Systems: New Directions in the Sociology and History of Technology*. Cambridge, MA: MIT Press, 1987.

Bill Budge's 3-D Graphics System and Game Tool. User manual. Davis, CA: California Pacific Computer Co., 1980. https://www.apple.asimov.net/documentation/applications/misc/Bill%20Budges%203-D%20Graphics%20System%20and%20Game%20Tool.pdf.

Birkel, Michael. "Leadings and Discernment." In *The Oxford Handbook of Quaker Studies*, edited by Stephen W. Angell and Ben Pink Dandelion, 245–59. Oxford: Oxford University Press, 2013.

Block, Ryan. "Live from Macworld 2007: Steve Jobs Keynote." *Engadget*, January 9, 2007. https://www.engadget.com/2007-01-09-live-from-macworld-2007-steve-jobs-keynote.html.

"Blumenthal Attacks Plan to Cut Tax Rate on Capital Gains, Wrangling Senate Panel." *Wall Street Journal*, June 29, 1978.

Boenig-Liptsin, Margarita. "Making Citizens of the Information Age: A Comparative Study of the First Computer Literacy Programs for Children in the United States, France, and the Soviet Union, 1970–1990." PhD diss., Harvard University, 2015.

Boston, Bruce O. "Search Series, Educational Software for TRS-80." *InfoWorld*, April 18, 1983, 54–56.

Bowman, Bill. Interview by Kay Savetz. *ANTIC: The Atari 8-bit Podcast*. May 16, 2017. https://ataripodcast.libsyn.com/antic-interview-278-bill-bowman-ceo-of-spinnaker-software.

Brand, Stewart. "SPACEWAR: Fantastic Life and Symbolic Death among the Computer Bums." *Rolling Stone*, December 7, 1972.

Braun, Joseph A., Jr. *Microcomputers and the Social Studies: A Resource Guide for the Middle and Secondary Grades*. New York: Garland, 1986. https://archive.org/details/microcomputerss0000brau/.

Bricklin, Dan. *Bricklin on Technology*. New York: Wiley, 2009.

———. "Special Short Paper for the HBS Advertising Course." 1978. http://www.bricklin.com/anonymous/bricklin-1978-visicalc-paper.pdf.

Bricklin, Dan, and Bob Frankston. Interview by Martin Campbell-Kelly and Paul Ceruzzi, May 7, 2004. Charles Babbage Institute. https://conservancy.umn.edu/bitstream/handle/11299/113026/oh402b%26f.pdf.

Brøderbund Software, Inc., Collection. Brian Sutton-Smith Library and Archives of Play, Strong National Museum of Play, Rochester, NY.

Butcher, Lee. *Accidental Millionaire: The Rise and Fall of Steve Jobs at Apple Computer*. New York: Paragon House, 1988.

Cain, Victoria. *Schools and Screens: A Watchful History*. Cambridge, MA: MIT Press, 2021.

California State Board of Education. *Program Descriptions for History-Social Science Instruction Materials*. Sacramento: California State Board of Education, 1983. https://archive.org/details/ERIC_ED240019/.

Campbell, Gregory D. Letter to the editor. *BYTE*, September 1975.

Campbell-Kelly, Martin. *From Airline Reservations to Sonic the Hedgehog: A History of the Software Industry*. Cambridge, MA: MIT Press, 2003.

———. "Number Crunching without Programming: The Evolution of Spreadsheet Usability." *IEEE Annals of the History of Computing* 29, no. 3 (July–September 2007): 6–9.

Campbell-Kelly, Martin, William Aspray, Nathan Ensmenger, and Jeffrey R. Yost. *Computer: A History of the Information Machine*. 3rd ed. Boulder, CO: Westview Press, 2014.

Carlston, Doug. Foreword to *The Official Print Shop Handbook: Ideas, Tips, and Designs for Home, School, and Professional Use*, by Randi Benton and Mary Schenck Balcer, xiii. New York: Bantam Books, 1987. https://archive.org/details/officialprintshoooobent/.

———. Interview by Tim Bergin, November 19, 2004, Mountain View, CA. Transcript, Computer History Museum. http://archive.computerhistory.org/resources/text/Oral_History/Carlston_Doug/Carlston_Doug_1.oral_history.2004.102658043.pdf.

———. *Software People: Inside the Computer Business*. New York: Simon & Schuster, 1985.

Carpenter, Chuck. "Apple II vs Apple II Plus." *Creative Computing*, May 1980.

Carroll, Peter N. *It Seemed Like Nothing Happened: America in the 1970s*. New Brunswick, NJ: Rutgers University Press, 1990.

Carter, Jimmy. "Energy and the National Goals: Address to the Nation." Delivered July 15, 1979. Jimmy Carter Presidential Library and Museum. https://www.jimmycarterlibrary.gov/assets/documents/speeches/energy-crisis.phtml.

Caruso, Denise. "Software Gambles: Company Strategies Boomerang." *InfoWorld*, April 2, 1984, 80–83.

———. "VisiCorp, Software Arts Battle for *VisiCalc* Rights." *InfoWorld*, March 5, 1984, 13–15.

Cass, Stephen. "Al Alcorn, Creator of *Pong*, Explains How Early Home Computers Owe Their Color Graphics to This One Cheap, Sleazy Trick." *IEEE Spectrum*, April 21, 2020. https://spectrum.ieee.org/tech-talk/tech-history/silicon-revolution/al-alcorn-creator-of-pong-explains-how-early-home-computers-owe-their-color-to-this-one-cheap-sleazy-trick.

CBS. "The CBS Video Game Plan." Print advertisement. *Billboard*, December 11, 1982, 2–3.

CBS News. "CBS News Election Coverage: November 4, 1952." NewsActive3: A Television News Archive. YouTube video, 31:02. https://www.youtube.com/watch?v=5vjDod8D9Ec.

Ceruzzi, Paul E. *A History of Modern Computing*. 2nd ed. Cambridge, MA: MIT Press, 2003.

Chamberlain, Steve. "Apple II Copy Protection." *Big Mess o' Wires* (blog), August 27, 2015. https://www.bigmessowires.com/2015/08/27/apple-ii-copy-protection/.

Chion-Kenney, Linda. "Schools Bought Record Number of Computers in 1984." *Education Week*, March 27, 1985.

Ciarcia, Steve. "Computerize a Home." *BYTE*, January 1980, 28–54.

Coburn, Edward J. *Learning about Microcomputers: Hardware and Application Software*. Albany, NY: Delmar, 1986.

Cohen, Scott. *Zap: The Rise and Fall of Atari*. New York: McGraw-Hill, 1984.

Cohl, Claudia. "Join Us in Family Computing." *Family Computing*, September 1983.

Computer Faire. Program for the First West Coast Computer Faire, April 15–17, 1977. https://archive.org/details/the_First_West_Coast_Computer_Faire_April-1977/.

Con Díaz, Gerardo. *Software Rights: How Patent Law Transformed Software Development in America*. New Haven, CT: Yale University Press, 2019.

"Congress and Estate Taxes . . . Venture Capital's Record Year." *Wall Street Journal*, June 22, 1981.

Conniffe, Patricia. *Computer Dictionary*. New York: Scholastic, 1984.

Coughlin, Tom. "Chuck Peddle—Creator of the MOS 6502 Microprocessor Passes." *IEEE Consumer Electronics Magazine* 9, no. 3 (May 2020): 6–7.

CRAW. "Teri Perl." Women in Computer Science, Computing Research Association Committee on the Status of Women in Computing Research (CRAW). Accessed December 31, 2021. http://users.sdsc.edu/~jsale/CRAW/craw_bro.html.

Crawford, Chris. *The Art of Computer Game Design: Reflections of a Master Game Designer*. Berkeley, CA: Osborn/McGraw-Hill, 1984. https://archive.org/details/artofcomputergam00chri/.

Cringely, Robert X. [Mark Stephens]. *Accidental Empires: How the Boys of Silicon Valley Make Their Millions, Battle Foreign Competition, and Still Can't Get a Date*. New York: Harper, 1996.

———. "Verizon's iPhone Story Isn't So Black and White." *I, Cringely* (blog), January 11, 2011. https://www.cringely.com/2011/01/11/verizons-iphone-story-isnt-so-black-and-white/.

Cronk, Loren, and Richard Whittaker. *The Print Shop Reference Manual for the Apple*. n.p.: Brøderbund Software, 1984.

Cuban, Larry. *Oversold and Underused: Computers in the Classroom*. Cambridge, MA: Harvard University Press, 2001.

Digital Equipment Computer Users Society (DECUS). *DECUS Program Library Catalog*. Maynard, MA: Digital Equipment Computer Users Society, November 1969. https://ia801901.us.archive.org/23/items/bitsavers_decdecuspratalogNov69_6809707/DECUS_Catalog_Nov69.pdf.

———. *DECUS Program Library: PDP-11 Catalog*. Maynard, MA: Digital Equipment Corporation, August 1978. http://www.bitsavers.org/pdf/dec/decus/programCatalogs/DECUS_Catalog_PDP-11_Aug78.pdf.

———. *PDP-8 Program Library Catalog*. Maynard, MA: Digital Equipment Computer Users Society, June 1979. https://archive.org/details/bitsavers_decdecusprDP8ProgramLibraryCatalogJun79_14606795/.

Digital Equipment Corporation. *Microcomputer Handbook*. Maynard, MA:

Digital Equipment Corporation, 1976. https://archive.org/details/bitsavers_decpdp11harHandbook19761977_34390691.

"Digital Research Attracts Investors: Maker of CP/M Expands." *InfoWorld*, October 5, 1981.

Donelan, B. L. Letter to the editor. *BYTE*, June 1976.

Donovan, Tristan. *Replay: The History of Video Games*. East Sussex: Yellow Ant, 2010.

Driscoll, Kevin. "Professional Work for Nothing: Software Commercialization and 'An Open Letter to Hobbyists.'" *Information & Culture* 50, no. 2 (2015): 257–83.

Edwards, Paul N. *The Closed World: Computers and the Politics of Discourse in Cold War America*. Cambridge, MA: MIT Press, 1996.

Eisenhart, Mary. "Broderbund Software: Next Stop World Domination?" *Microtimes*, September 1985, 36–42.

Ekblaw, Richard. Letter to the editor. *Softalk*, April 1983.

"Electronic Computer for Home Operation (ECHO): The First Home Computer." *IEEE Annals of the History of Computing* 16, no. 3 (Fall 1994): 59–61.

Electronic Data Systems Corporation. *Little Computers . . . See How They Run*. Dallas, TX: Electronic Data Systems Corporation, 1980.

Ensmenger, Nathan L. *The Computer Boys Take Over: Computers, Programmers, and the Politics of Technical Expertise*. Cambridge, MA: MIT Press, 2012.

Faggin, Frederico, Marcian E. Hoff Jr., Stanley Mazor, and Masatoshi Shima. "The History of the 4004." *IEEE Micro* 16, no. 6 (December 1996): 10–20.

Felsenstein, Lee. "Excerpts from a Chalk Talk on the Tom Swift Terminal by Lee Felsenstein Presented at the June 11, 1975, HCC Meeting." Homebrew Computer Club newsletter, July 5, 1975.

Fields III, J. Burford. Letter to the editor. *Softalk*, August 1982.

Finnigan, David. *The New Apple II User's Guide*. Lincoln, IL: Mac GUI, 2012.

Fiske, Edward B. "Schools Enter the Computer Age: An Analysis." *New York Times*, April 25, 1982, sec. 12.

Fitzpatrick, Karen. "Bit Copy Programs." *Hardcore Computing* 1, no. 1 (1981): 10–13.

Ford, Clyde W. *Think Black: A Memoir*. New York: Amistad, 2019.

Fox, Joy, and Richard Fox. "Biorhythm for Computers." *BYTE*, April 1976, 20–23.

Freiberger, Paul, and Michael Swaine. *Fire in the Valley: The Making of the Personal Computer*. New York: Osborne/McGraw-Hill, 1984.

Freiberger, Stephen, and Paul Chew. *A Consumer's Guide to Personal Computing and Microcomputers*. Rochelle Park, NJ: Hayden Book Company, 1978.

Friedberg, Anne. *The Virtual Window: From Alberti to Microsoft*. Cambridge, MA: MIT Press, 2009.

Fylstra, Dan. "Personal Account: The Creation and Destruction of VisiCalc." May 2004. https://www.computerhistory.org/collections/catalog/102738286.

———. "The Radio Shack TRS-80: An Owner's Report." *BYTE*, April 1978.

———. "User's Report: The PET 2001." *BYTE*, March 1978.

Fylstra, Dan, and Jef Raskin. *Microchess 2.0*. User manual. n.p.: Personal Software Inc., 1978. https://mirrors.apple2.org.za/Apple%20II%20Documentation%20Project/Software/Cassettes/Microchess%202.0%20Manual.pdf.

Gaboury, Jacob. "Screen Selfies and High Scores." Fotomuseum Witherthur, May 7, 2019, https://www.fotomuseum.ch/en/2019/07/05/screen-selfies-and-high-scores/.

Gardner, Richard. "The Shadow, Buck Rogers, and the Home Computer." *BYTE*, October 1975, 58–60.

Gates, Bill. "An Open Letter to Hobbyists." Homebrew Computer Club newsletter, January 31, 1976.

———. "Software Contest Winners Announced." *Computer Notes*, Altair Users Group, July 1975.

Gear, Tommy. "Backtalk." *Softalk*, January 1984.

Goehner, Ken. "Pixellite Software: The People Behind The Print Shop." *Microtimes*, April 1985, 30–43. https://archive.org/details/microtimesvolume00bamp_6/.

Goldberg, Marty, and Curt Vendel. *Atari, Inc.: Business Is Fun*. Carmel, NY: Syzygy Company Press, 2012.

Golding, Val. "Rebuttal: A Letter from: Val Golding." *Hardcore Computing* 1, no. 2 (1981): 12. https://archive.org/details/hardcore-computing-2/.

Gotkin, Kevin. "When Computers Were Amateur." *IEEE Annals of the History of Computing* 36, no. 2 (April–June 2014): 4–14.

Grad, Burton. "The Creation and the Demise of VisiCalc." *IEEE Annals of the History of Computing* 29, no. 3 (July–September 2007): 20–31.

Grady, M. Tim, and Jane D. Gawronski. *Computers in Curriculum and Instruction*. Alexandria, VA: Association for Supervision and Curriculum Development, 1983.

Green, Wayne. "Painful Facts of Life." *80 Microcomputing*, August 1980.

———. "Remarks from the Publisher . . ." *inCider*, January 1983.

Guins, Raiford. *Atari Design: Impressions on Coin-Operated Video Game Machines*. New York: Bloomsbury, 2020.

———. *Game After: A Cultural Study of Video Game Afterlife*. Cambridge, MA: MIT Press, 2014.

Gutman, Dan. "Praising Weirdware." *InfoWorld*, November 26, 1984.

Haigh, Thomas, and Paul E. Ceruzzi. *A New History of Modern Computing*. Cambridge, MA: MIT Press, 2021.

Haight, Bev R. "Censorship in Computer Magazines." *Hardcore Computing* 1, no. 1 (1981): 4–5. https://archive.org/details/hardcore-computing-1/.

Haight, Chuck R. "What I Need Is a USER's Magazine." *Hardcore Computing* 1, no. 1 (1981): 2. https://archive.org/details/hardcore-computing-1/.

Hallen, Rod. "INDXA: A BASIC Routine File Index." *Creative Computing*, November/December 1978.

Haller, George. "Golf Handicapping." *BYTE*, February 1976, 46–47.

Haring, Kristen. *Ham Radio's Technical Cultures*. Cambridge, MA: MIT Press, 2007.

Harman, Jill B. Letter to the editor. *Softalk*, May 1984. https://archive.org/details/softalkv4n09may1984/.

Harrington, William H. "Marketalk Reviews." *Softalk*, May 1983.

Harvey, David. *A Brief History of Neoliberalism*. New York: Oxford University Press, 2005.

Helmers, Carl. "Magnetic Recording Technology." *BYTE*, March 1976.

———. "Returning to the Tower of Babel, or . . . Some Notes about LISP, Languages and Other Topics . . ." *BYTE*, August 1979.

———. "What Is BYTE—(the First) Editorial." *BYTE*, September 1975.

Hertzfeld, Andy. "Reality Distortion Field." *The Original Macintosh* (blog), *Folklore*. https://www.folklore.org/StoryView.py?story=Reality_Distortion_Field.txt.

Hixson, Amanda. "Lotus 1-2-3, Integrated Program for the IBM PC." *InfoWorld*, August 1, 1983, 39–41.

Hodges, James. "Technical Fixes for Legal Uncertainty in the 1980s Software Cracking Scene." *IEEE Annals of the History of Computing* 41, no. 4 (2019): 20–33.

Hoeppner, John. "Interface a Floppy-Disk Drive to a 8080A-Based Computer." *BYTE*, May 1980.

Honan, Mathew. "Apple Unveils iPhone." MacWorld, January 9, 2007. https://www.macworld.com/article/1054769/iphone.html.

Hunter, Beverly. *My Students Use Computers: Computers Literacy in the K–8 Curriculum*. Reston, VA: Reston Publishing Company, 1984. https://archive.org/details/mystudentsusecom0000unse/.

Hurlocker, Charles. Letter to the editor. *BYTE*, June 1976.

Infield, Glenn. "A Computer in the Basement?" *Popular Mechanics*, April 1968.

"Inside the Industry." *Computer Gaming World*, April 1988.

International Resource Development. *Microcomputer Software Packages*. Report #569. Norwalk, CT: International Resource Development Inc., September 1983. https://archive.org/details/1983MicroSoftwarePackagesIntnlRsourceDev/.

Isaacson, Walter. *The Innovators: How a Group of Hackers, Geniuses, and Geeks Created the Digital Revolution*. New York: Simon & Schuster, 2014.

———. *Steve Jobs: The Exclusive Biography*. New York: Simon & Schuster, 2011.

"It's *Choplifter* in '82 and *Wizardry* for All-Time Pops." *Softalk*, April 1983.

JCPenney. *Christmas 1980*. https://christmas.musetechnical.com/ShowCatalog/1980-JCPenney-Christmas-Book.

Jennings, Peter. Interview by Sellam Ismail, February 1, 2005. "Oral History of Peter Jennings." Computer History Museum. http://archive.computerhistory.org/projects/chess/related_materials/oral-history/jennings.oral_history.2005.102630656/jennings.oral_history_transcrit.2005.102630656.pdf.

———. *MICROCHESS*. User manual. Toronto: Micro-Ware Ltd., 1976. Reproduced in raw text format on Benlo Park (Peter Jennings's personal website). http://www.benlo.com/microchess/Kim-1Microchess.html.

———. *MicroChess for the KIM-1*. User manual. Chelmsford, MA: The Computerist, 1977. http://retro.hansotten.nl/uploads/files/microchess%20manual.pdf.

———. "VisiCalc—The Early Days." Benlo Park (Peter Jennings's personal website). http://www.benlo.com/visicalc/index.html.

———. "VisiCalc 1979," 2. Benlo Park. http://www.benlo.com/visicalc/visicalc2.html.

———. "VisiCalc 1979," 3. Benlo Park. http://www.benlo.com/visicalc/visicalc3.html.

Jobs, Steve. "iPhone 1—Steve Jobs MacWorld Keynote in 2007—Full Presentation, 80 mins." Protectstar Inc. Recorded on January 9, 2007. YouTube video, 1:19:10. https://www.youtube.com/watch?v=VQKMoT-6XSg.

Johns, Adrian. *Piracy: The Intellectual Property Wars from Gutenberg to Gates*. Chicago: University of Chicago Press, 2010.

Johnson, Ethan, and Alex Smith. "A Breakout Story." *The History of How We Play* (blog), December 29, 2018. https://thehistoryofhowweplay.wordpress.com/2018/12/29/a-breakout-story/.

Kelty, Christopher. "Inventing Copyleft." In *Making and Unmaking Intellectual Property: Creative Production in Legal and Cultural Perspective*, edited by Mario Bagioli, Peter Jaszi, and Martha Woodmansee, 133–48. Chicago: University of Chicago Press, 2011.

Kilby, Jack S. "The Integrated Circuit's Early History." *Proceedings of the IEEE* 88, no. 1 (2000): 109–11.

Kirschenbaum, Matthew. *Mechanisms: New Media and the Forensic Imagination*. Cambridge, MA: MIT Press, 2012.

———. *Track Changes: A Literary History of Word Processing*. Cambridge, MA: Belknap Press of Harvard University Press, 2016.

Kline, Stephen, Nick Dyer-Witheford, and Greig de Peuter. *Digital Play: The Interaction of Technology, Culture and Marketing*. Montreal: McGill-Queen's University Press, 2003.

Kneale, Dennis. "Overloaded System: As Software Products and Firms Proliferate, a Shakeout Is Forecast." *Wall Street Journal*, February 23, 1984.

Kocurek, Carly. *Coin-Operated Americans: Rebooting Boyhood at the Video Game Arcade*. Minneapolis: University of Minnesota Press, 2015.

Kortum, Sarah. "Confessions of a Reformed Computer Phobic." *Family Computing*, September 1983, 34–37.

Lancaster, Don. "TV Typewriter." *Radio-Electronics*, September 1973.

Larson, Erik. "Many Firms Seek Entry into Software." *Wall Street Journal*, January 6, 1984.

Lau, Ted M. "Total Kitchen Information System." *BYTE*, January 1976, 42–45.

Leavitt, Michael R. Letter to the editor. *InfoWorld*, July 6, 1981.

Levering, Robert, Michael Katz, and Milton Moskowitz. *The Computer Entrepreneurs: Who's Making It Big and How in America's Upstart Industry*. New York: New American Library, 1984.

Levine, James A., Joan Levine, Jessica Levine, and Joshua Levine. "A Family Computer Diary." In *Digital Deli: The Comprehensive, User-Lovable Menu of Computer Lore, Culture, Lifestyles and Fancy*, edited by Steve Ditlea, 117–19. New York: Workman Publishing, 1984. Republished online at AtariArchives.org, https://www.atariarchives.org/deli/wall_street.php. Originally published in *Parents* magazine, July 1983.

Levy, Steven. *Hackers: Heroes of the Computer Revolution*. New York: Penguin, 2001. First published Garden City, NY: Anchor Press/Doubleday, 1984.

———. *Insanely Great: The Life and Times of Macintosh, the Computer that Changed Everything*. New York: Penguin Books, 1994.

———. “A Spreadsheet Way of Knowledge.” *Harper’s*, November 1984. Reposted on *Wired*, October 24, 2014. https://www.wired.com/2014/10/a-spreadsheet-way-of-knowledge/.

Lewenstein, Bruce V. “The Ethics of Software Piracy.” *PC Magazine*, April 30, 1985.

Litterick, Ian. *How Computers Work*. New York: Bookwright Press, 1984.

Lock, Robert. “The Editor’s Notes.” *COMPUTE!* March 1981.

———. “Editor’s Notes.” *COMPUTE!*, November 1983.

Lojek, Bo. *A History of Semiconductor Engineering*. Berlin: Springer, 2007.

Lubar, David. “Educational Software: Part One.” *Creative Computing*, September 1980, 64–72.

———. “Software, Hardware, and Otherware for Christmas.” *Creative Computing*, December 1980.

Malcolm, Michael A., and Gary R. Sager, “The Real-Time/Minicomputer Laboratory,” University of Waterloo, February 1976. https://cs.uwaterloo.ca/research/tr/1976/CS-76-11.pdf.

Malone, Michael S. *Infinite Loop: How the World’s Most Insanely Great Computer Company Went Insane*. New York: Doubleday Business, 1999.

Manes, Stephen, and Paul Andrews. *Gates: How Microsoft’s Mogul Reinvented an Industry and Made Himself the Richest Man in America*. New York: Doubleday, 1992.

Marcus, Stanley. *His and Hers: The Fantasy World of the Neiman-Marcus Catalogue*. New York: Viking Press, 1982.

Markoff, John. *What the Dormouse Said: How the Sixties Counterculture Shaped the Personal Computer Industry*. New York: Viking Penguin, 2005.

Maxis. “Maxis Print Artist Advertisement (1995).” Andy C (YouTube user). YouTube video, 1:27. https://www.youtube.com/watch?v=irAn64mRcyY.

Mazur, Jeffrey. “Hardtalk.” *Softalk*, January 1983. https://archive.org/details/softalkv3n05jan1983/.

McCollough, Karen G. “The Print Shop for Apple, Atari, and Commodore 64.” *COMPUTE!*, March 1985, 74–76.

McFadden, Andrew. “Early Copy Protection on the Apple II.” Fadden

.com (personal website). Accessed July 31, 2021. https://fadden.com/apple2/cassette-protect.html.

McIlwain, Charlton. *Black Software: The Internet and Racial Justice, from the AfroNet to Black Lives Matter*. New York: Oxford University Press, 2019.

McKenna, Regis. *The Regis Touch: New Marketing Strategies for Uncertain Times*. New York: Addison-Wesley, 1985.

McMullen, Barbara E., and John F. McMullen. "Screen Envy on Wall Street." In *Digital Deli: The Comprehensive, User-Lovable Menu of Computer Lore, Culture, Lifestyles and Fancy*, edited by Steve Ditlea, 277–78. New York: Workman, 1984. Republished online at AtariArchives.org, https://www.atariarchives.org/deli/wall_street.php.

McWilliams, Peter A. *The Personal Computer Book*. Los Angeles: Prelude Press, 1982.

Meehan, Michael. Letter to the editor. *Softalk*, October 1982.

"Member Microcomputer Systems." Homebrew Computer Club newsletter, October 31, 1975.

"Memorial: Jere Hutchins Dykema '48." *Princeton Alumni Weekly*, February 9, 1994, 40. https://paw.princeton.edu/memorial/jere-hutchins-dykema-%E2%80%9948.

Metz, Robert. "Pac-Man and Beyond." Market Place. *New York Times*, June 4, 1982, D6, National ed. https://www.nytimes.com/1982/06/04/business/market-place-pac-man-and-beyond.html.

"Micro Use Is Up 75% School Survey Shows." *School Library Journal* 32 (December 1985): 12.

Mihm, Mickey T. "Software Piracy and the Personal Computer: Is the 1980 Software Copyright Act Effective?" *John Marshall Journal of Information Technology & Privacy Law* 4, no. 1 (1983): 171–93. https://repository.law.uic.edu/cgi/viewcontent.cgi?article=1553&context=jitpl.

Mikkelson, David. "Ken Olsen." *Snopes*, September 21, 2004. https://www.snopes.com/fact-check/ken-olsen/.

Milewski, Richard. "VCOPY Clones VisiCalc." *InfoWorld*, December 8, 1980.

Misa, Thomas J. "Military Needs, Commercial Realities, and the Development of the Transistor, 1948–1958." In *Military Enterprise and Technological Change*, edited by Merritt Roe Smith, 253–87. Cambridge, MA: MIT Press, 1985.

MITS. Advertisement. *Digital Design*, June 1975. New York Public Library, call nos. ZAN-V1546 and ZAN-V561. Photocopy available online at http://altair.ftldesign.com/.

———. *The Age of Altair*. Albuquerque, NM: MITS, 1975. https://archive.org/details/bitsavers_mits8800Allog1975_2964267.

———. *The Altair System*. Albuquerque, NM: MITS, 1976. https://archive.org/details/MITS_MITS_TheAltairSystem_Brochure/.

———. "MITS-mas: Special Altair Christmas Catalog." MITS, 1975. https://archive.org/details/altairchristmascatalog/.

"Monthly Computer Census." *Computers and Automation*, August 1968, 66–68.

Moore, Gordon. "Cramming More Components onto Integrated Circuits." *Electronics*, August 19, 1965.

Morgan, Chris. "Editorial: How Can We Stop Software Piracy?" *BYTE*, May 1981.

———. "Over into the Future." *Popular Computing*, November 1981.

Moritz, Michael. *The Little Kingdom: The Private Story of Apple Computer*. New York: William Morrow, 1984.

"The Most Popular Program of 1981." *Softalk*, April 1982.

Mueller, Gavin. *Media Piracy in the Cultural Economy: Intellectual Property and Labor under Neoliberal Restructuring*. New York: Routledge, 2019.

Naritomi, E. K. Letter to the editor. *InfoWorld*, June 8, 1981.

National Commission on Excellence in Education. *A Nation at Risk: The Imperative for Educational Reform*. April 1983.

Neiburger, E. J. Letter to the editor. *BYTE*, June 1982.

The Neiman-Marcus Christmas Book. Neiman-Marcus, 1969. Select pages available in the Computer History Museum's digital archives, http://archive.computerhistory.org/resources/access/text/2011/01/102685475-03-01-acc.pdf.

Nelson, Theodor H. *Computer Lib/Dream Machines*. n.p.: Self-published, 1974.

———. *The Home Computer Revolution*. South Bend, IN: Self-published, 1977.

"New Charts for Computers, Vidisks." *Billboard*, October 8, 1983.

Newman, Michael Z. *Atari Age: The Emergence of Video Games in America*. Cambridge, MA: MIT Press, 2017.

Nicholas, Tom. *VC: An American History*. Cambridge, MA: Harvard University Press, 2019.

"No More!" Editorial. *InfoWorld*, April 13, 1981.

Nooney, Laine. "Let's Begin Again: Sierra On-Line and the Origins of the Graphical Adventure Game." *American Journal of Play* 10, no. 1 (Fall 2017): 71–98.

———. "The Uncredited: Work, Women, and the Making of the U.S. Computer Game Industry." *Feminist Media Histories* 6, no. 1 (2020): 119–46. https://doi.org/10.1525/fmh.2020.6.1.119.

Nooney, Laine, Kevin Driscoll, and Kera Allen. "From Programming to Products: *Softalk* Magazine and the Rise of the Personal Computer User." *Information & Culture* 55, no. 2 (2020): 105–29.

North, Steve. "Apple II Computer." *Creative Computing*, July–August 1978.

Noyce, Robert N., and Marcian E. Hoff Jr. "A History of Microprocessor Development at Intel." *IEEE Micro* 1, no. 1 (February 1981): 8–21.

O'Donnell, Casey. "Production Protection to Copy(right) Protection: From the 10NES to DVDs." *IEEE Annals of the History of Computing* 31, no. 3 (2009): 54–63.

Olson, Gary. "A Family Computer Album." *Saturday Evening Post*, April 1982, 70–77.

Omega Software Systems. "*Locksmith*: Apple Disk Copy." Print advertisement. *MICRO: The 6502 Journal*, January 1981, 80. https://archive.org/details/micro-6502-journal-32/.

The Original Locksmith Users Manual: Version 5.0. Chicago: Omega MicroWare, 1983. https://archive.org/details/stx_Omega_Microware_Original_Locksmith_5.0_manual/.

Orme, Michael. *Micros: A Pervasive Force*. London: Associated Business Press, 1979.

Pelczarski, Mike. Letter to the editor. *BYTE*, June 1982.

Personal Software. "VisiCalc." Sunnyvale, CA: Personal Software,

[1980?]. https://archive.org/details/TNM_VisiCalc_-_Personal_Software_Inc_20170922_0447/.

Petrick, Elizabeth. "Imagining the Personal Computer: Conceptualizations of the Homebrew Computer Club 1975–1977." *IEEE Annals of the History of Computing* 39, no. 4 (2017): 27–39.

Pierson, John. "Steiger's Capital Gains Steamroller." *Wall Street Journal*, May 31, 1978.

Pollack, Andrew. "How a Software Winner Went Sour." *New York Times*, February 26, 1984, F1–F12.

———. "Visicorp Is Merging into Paladin." *InfoWorld*, November 3, 1984, 29.

"The Programs You Like Best." *Softalk*, April 1983. https://archive.org/details/softalkv3n08apr1983/.

Radio Shack. *Expansion Interface: Catalog Number 26-1140/1141/1142*. TRS-80 Micro Computer System user manual. Fort Worth, TX: Radio Shack, 1979. https://wiki.theretrowagon.com/w/images/c/c8/Radio_Shack_Expansion_Interface_Manual.pdf.

———. *1980 Catalog*. Fort Worth, TX: Tandy Corporation, 1979. https://www.radioshackcatalogs.com/flipbook/1980_radioshack_catalog.html.

Ramirez, Renya K. *Native Hubs: Culture, Community, and Belonging in Silicon Valley and Beyond*. Durham, NC: Duke University Press, 2007.

Rankin, Joy Lisi. *A People's History of Computing in the United States*. Cambridge, MA: Harvard University Press, 2018.

Reed, Lori. "Domesticating the Personal Computer: The Mainstreaming of a New Technology and the Cultural Management of a Widespread Technophobia, 1964–." *Critical Studies in Media Communication* 17, no. 2 (2000): 159–85.

Reed, Sally. "Schools Enter the Computer Age." *New York Times*, April 25, 1982, sec. 12.

Roberts, Ed, and William Yates. "Altair 8800 Minicomputer, Part 1." *Popular Electronics*, January 1975.

Rosch, Winn. "Printer Comparisons: Getting It on Paper with an Apple." *A+ Magazine*, November 1983. https://archive.org/details/aplus-v1-n01/.

Rose, Frank. *West of Eden: The End of Innocence at Apple Computer*. New York: Penguin, 1989.

Rosen, Benjamin M. "VisiCalc: Breaking the Personal Computer Software Bottleneck." *Morgan Stanley Electronics Letter*, Morgan Stanley, July 11, 1979.

Rosenberg, Ronald. "*A* Is for Acquisition: Watertown Educational Software, TV Production Firm Sold." *Boston Globe*, February 27, 1996.

———. "Bowman Leaves Spinnaker Software." *Boston Globe*, November 28, 1986.

Rosenthal, Peter. "Atari Speaks Out." Interview by Dave Ahl. *Creative Computing*, August 1979. https://archive.org/details/creativecomputing-1979-08/.

Rosner, Richard. "It's More Fun than Crayons." *BYTE*, November 1976.

Salsberg, Art. Editorial. *Popular Electronics*, January 1975.

"Scholastic Acquires Tom Snyder Productions from Canadian Publisher, Torstar Corp." *PRNewswire*, December 21, 2001.

Schuyler, Jim. "DesignWare's Founding." *Sky's Blog* (Schuyler's personal blog). September 19, 2017. https://blog.red7.com/designware-founding/.

———. "Jim Schuyler, Founder of DesignWare." Interview by Kay Savetz, September 11, 2017. YouTube video, 1:42:26. https://www.youtube.com/watch?v=oBW-z-C5X8A.

"The Shakeout in Software. It's Already Here." *Businessweek*, August 20, 1984, 102–4.

Shelly, Gary B., and Thomas J. Cashman. *Computer Fundamentals for an Information Age*. Brea, CA: Anaheim Publishing Company, 1984.

Shetterly, Margot Lee. *Hidden Figures: The Untold Story of the African American Women Who Helped Win the Space Race*. New York: HarperCollins, 2016.

"Showgirls, Splash, Finesse, Not Much Else: Summer CES Wows the Crowd with Glitter and Big Talk." *Softalk*, August 1983.

Sigel, Efrem, and Louis Giglio. *Guide to Software Publishing: An Industry Emerges*. White Plains, NY: Knowledge Industry Publications, 1984.

Sipe, Russell. "Computer Games in 1983: A Report." *Computer Gaming World*, March 1983.

Sirotek, Robert. Letter to the editor. *Softalk*, August 1982.

Sklarewitz, Norman. "Born to Grow." *Inc.*, April 1979.

"Software Prices . . ." *Computer Notes*, April 7, 1975, 3. https://archive.org/details/Computer_Notes_1975_01_01/.

Software Publishers Association. "An Open Letter to the User Community." *COMPUTE!*, March 1985.

Southwestern Data Systems. "Apple-Doc by Roger Wagner." Product insert. Southwestern Data Systems, Santee, CA, 1981. Viewed on an eBay listing posted by user "klr-store" (Kevin Lefloch). https://web.archive.org/web/20211108202009/https://www.ebay.com/itm/SDS-Southwestern-Data-Systems-Apple-Doc-1979-By-Roger-Wagner-Apple-II-vintage-/121562547579.

Spicer, Dag. "The ECHO IV Home Computer: 50 Years Later." Computer History Museum blog, May 31, 2016. https://computerhistory.org/blog/the-echo-iv-home-computer-50-years-later/.

———. "If You Can't Stand the Coding, Stay Out of the Kitchen: Three Chapters in the History of Home Automation." *Dr. Dobb's*, August 12, 2000. https://www.drdobbs.com/architecture-and-design/if-you-cant-stand-the-coding-stay-out-of/184404040.

Spinnaker Software Corporation. Harvard Business School case study 9-385-252. Prepared by José-Carlos Jarillo Mossi. Boston, MA: Harvard Business Publishing, 1985.

Staples, Betsy. "Educational Computing: Where Are We Now?" *Creative Computing*, April 1985.

Stein, Jesse Adams. "Domesticity, Gender, and the 1977 Apple II Personal Computer." *Design and Culture* 3, no. 2 (2011): 193–216.

Sudnow, David. *Pilgrim in the Microworld: Eye, Mind, and the Essence of Video Skill*. New York: Warner Books, 1983.

"A Summer-CES Report." *Boston Phoenix*, September 6, 1983, sec. 4.

Švelch, Jaroslav. *Gaming the Iron Curtain: How Teenagers and Amateurs in Communist Czechoslovakia Claimed the Medium of Computer Games*. Cambridge, MA: MIT Press, 2018.

Tanenbaum, Andrew S., Paul Klint, and Wim Bohm, "Guidelines for Software Portability." In *Software: Practice and Experience*, 681–98. New York: Wiley, 1978.

Taylor, Robert, ed. *The Computer in the School: Tutor, Tool, Tutee*. New York: Teachers College Press, 1980.

"Thanks." Editorial. *Hardcore Computing* 1, no. 2 (1981): 11. https://archive.org/details/hardcore-computing-2/.

Titus, Jon. "Build the Mark 8 Minicomputer." *Radio-Electronics*, July 1974, 29–33.

Toffler, Alvin. *Future Shock*. New York: Random House, 1970.

———. *The Third Wave*. New York: William Morrow, 1980.

Tommervik, Allan. "Assembling Useful Utilities." *Softalk*, August 1981.

———. "Exec Edu-Ware." *Softalk*, May 1981.

———. "Exec Personal: The VisiCalc People." *Softalk*, October 1980.

———. "Exec Softsel." *Softalk*, October 1981.

———. "The Great Arcade/Computer Controversy, Part One: The Publishers and the Pirates." *Softline*, January 1982.

———. "On-Line Exec: Adventures in Programming." *Softalk*, February 1981.

———. "The Staggering Value of Pirate's Booty." *Softalk*, October 1980. https://archive.org/details/softalkv1no02oct1980/.

Tommervik, Margot Comstock. "Marketalk Reviews." *Softalk*, September 1982.

———. "The Print Shop." *Softalk*, June 1984, 114–15.

———. "Straightalk." *Softalk*, September 1980.

Tripp, Robert. "Copyright/Copywrong." Editorial. *MICRO: The 6502 Journal*, March 1981.

Turner, Fred. *From Counterculture to Cyberculture: Stewart Brand, the Whole Earth Network, and the Rise of Digital Utopianism*. Chicago: University of Chicago Press, 2010.

"Unison Loses Software Suit." *New York Times*, October 16, 1986, D22.

Vacroux, Andre G. "Microcomputers." *Scientific American*, May 1975.

Varven, Jean. "The Schoolhouse Apple." *Softalk*, May 1982.

Veit, Stan. *Stan Veit's History of the Personal Computer*. Asheville, NC: Worldcomm, 1993.

"VisiCalc: User-Defined Problem Solving Package." *Intelligent Machines Journal* 1, no. 9 (June 11, 1979): 22. https://books.google.com/books?id=Gj4EAAAAMBAJ&pg=PA22.

Wallace, Bob. "The PET vs. the TRS-80." *MICRO: The 6502 Journal*, December 1977–January 1978.

Watt, Dan. "Computers in Education." *Popular Computing*, August 1983.

Watters, Audrey. *Teaching Machines: The History of Personalized Learning*. Cambridge, MA: MIT Press, 2021.

Weyhrich, Steven. *Sophistication and Simplicity: The Life and Times of the Apple II Computer*. Winnipeg: Variant Press, 2019.

Wilkes, M. V. *Time-Sharing Computer Systems*. London: Macdonald, 1968.

Williams, Dmitri. "Structure and Competition in the US Home Video Game Industry." *International Journal on Media Management* 4, no.1 (2002): 41–54.

Williams, Ken. "Computers: The Myth, the Promise, the Reason." *Creative Computing*, November 1984.

Willis, Jerry, and Merl Miller. *Computers for Everybody*. New York: Signet, 1983.

———. *Computers for Everybody*. Beaverton, OR: dilithium Press, 1984.

Willis, Jerry, with Deborrah Smithy and Brian Hyndman. *Peanut Butter and Jelly Guide to Computers*. Beaverton, OR: dilithium Press, 1978.

Wollman, Jane. "Software Piracy and Protection." *Popular Computing*, April 1982.

Woolgar, Stephen. "Configuring the User." In *A Sociology of Monsters: Essays on Power, Technology, and Domination*, edited by John Law, 57–99. New York: Routledge, 1991.

Worth, Don, and Pieter Lechner. *Beneath Apple DOS*. Reseda, CA: Quality Software, 1981.

Wozniak, Steve, with Gina Smith. *iWoz: Computer Geek to Cult Icon. How I Invented the Personal Computer, Co-Founded Apple, and Had Fun Doing It*. New York: Norton, 2006.

Yuen, Matthew T. "Head's Up! Exec Beagle Bros." *Softalk*, October 1983.

———. "Pirate, Thief. Who Dares to Catch Him?" *Softalk*, October 1980. https://archive.org/details/softalkv1n02oct1980/.

Zientara, Marguerite. "Five Powerful Women." *InfoWorld*, May 21, 1984.

———. *Women, Technology, and Power: Ten Stars and the History They Made.* New York: AMACOM, 1987.

Zonderman, Jon. "From Diapers to Disk Drives." *Family Computing*, September 1983, 30–32.

Zuckerman, Faye. "New on the Charts." *Billboard*, August 4, 1984, 31.

Index

Page numbers in italics indicate figures.